T0133377

Science without Frontiers

The OSU Press Horning Visiting Scholars
Publication Series

EDITORS: *Anita Guerrini and David Luft*

PREVIOUSLY PUBLISHED

Aetna and the Moon:

Explaining Nature in Ancient Greece and Rome

LIBA TAUB

Artisan/Practitioners and the Rise of the New Sciences, 1400–1600

PAMELA O. LONG

Grow Food, Cook Food, Share Food

KEN ALBALA

Science without Frontiers

Cosmopolitanism and National Interests in the World of Learning, 1870–1940

ROBERT FOX

Oregon State University Press | CORVALLIS

Library of Congress Cataloging-in-Publication Data

Names: Fox, Robert, 1938-

Title: Science without frontiers : cosmopolitanism and national interests in the world of learning, 1870–1940 / Robert Fox.

Description: Corvallis : Oregon State University Press, [2016] | Series: The OSU Press Horning Visiting Scholars publication series | Includes bibliographical references and index.

Identifiers: LCCN 2016035634 | ISBN 9780870718670 (original trade pbk. : alk. paper)

Subjects: LCSH: Communication in science—History—19th century. | Communication in science—History—20th century. | Nationalism and science—History—19th century. | Nationalism and science—History—20th century. | Science and state—History—19th century. | Science and state—History—20th century.

Classification: LCC Q223 .F698 2016 | DDC 338.9/2609041—dc23

LC record available at https://lccn.loc.gov/2016035634

♾ This paper meets the requirements of ANSI/NISO Z39.48-1992 (Permanence of Paper).

First published in 2016 by Oregon State University Press
Printed in the United States of America

Oregon State University Press
121 The Valley Library
Corvallis OR 97331-4501
541-737-3166 • fax 541-737-3170
www.osupress.oregonstate.edu

CONTENTS

ILLUSTRATIONS

ABBREVIATIONS

Libraries and archives

BL	British Library, London
BnF	Bibliothèque nationale de France, Paris
Bod.	Bodleian Library, University of Oxford
IUPAC Archives (CHF)	IUPAC Papers, Donald F. and Mildred Topp Othmer Library of Chemical History, Chemical Heritage Foundation, 315 Chestnut Street, Philadelphia, PA 19106, USA
Mundaneum Archives, Mons	Mundaneum Centre d'archives de la Fédération Wallonie-Bruxelles, Mons, Belgium

Institutional and other abbreviations

A number of the abbreviations have both French and English versions. In the text, I have not sought linguistic consistency but generally adopted the form that was most commonly used at the time.

BIPM	Bureau international des poids et mesures
CNR	Consiglio Nazionale delle Ricerche
CNRS	Centre national de la recherche scientifique
COIA	Central Office of International Associations (see also OCAI)
CSIC	Consejo superior de investigaciones científicas
DPLA	Digital Public Library of America
ETH	Eidgenössische Technische Hochschule, Zurich. Also commonly referred to as the Zurich Polytechnikum

EUR	Esposizione Universale di Roma, 1942 (also abbreviated as E42)
IAA	International Association of Academies
IACS	International Association of Chemical Societies
IAU	International Astronomical Union
ICIC	International Committee on Intellectual Co-operation
ICSL	*International Catalogue of Scientific Literature*
ICSU	International Council of Scientific Unions
IGA	International Geodetic Association
IIB	Institut international de bibliographie
IMU	International Mathematical Union
IRC	International Research Council
IUGG	International Union of Geodesy and Geophysics
IUPAC	International Union of Pure and Applied Chemistry
JAE	Junta para Ampliación de Estudios e Investigaciones Científicas
KWU	Kaiser-Wilhelms-Universität, Strasbourg
OCAI	Office central des associations internationales (see also COIA)
OCI	Office de coopération intellectuelle
OIB	Office international de bibliographie
UAI	Union des associations internationales (see also UIA)
UIA	Union of International Associations (see also UAI)
UDC	Universal Decimal Classification
UNESCO	United Nations Educational, Scientific, and Cultural Organization
URSI	Union radio-scientifique internationale (also International Union of Radio Science)

FOREWORD

Benjamin Horning (1890–1991) made the bequest, in memory of his parents, that established the Thomas Hart and Mary Jones Horning Endowment in the Humanities at Oregon State University. Mary Jones and Thomas Hart Horning were members of pioneering families of Benton County and Corvallis, Oregon. Benjamin Graham Horning graduated in 1914 from what was then Oregon Agricultural College and went on to complete a medical degree at Harvard and a degree in public health at Johns Hopkins University. Dr. Horning's long professional career included service in public health in Connecticut, work on rural health as a staff member with the American Public Health Association, and a position as medical director for the W. K. Kellogg Foundation, which led to his spending many years in Latin America. Dr. Horning wanted his bequest at Oregon State University to expand education in the humanities and to build a bridge between the humanities and the sciences.

Since 1994, the endowment has supported an annual lecture series and individual lectures, conferences, symposia, and colloquia, as well as teaching, research, and library collection development. The Horning professorships are housed in the Department of History in the School of History, Philosophy, and Religion. The first Thomas Hart and Mary Jones Horning Professors in the Humanities, Mary Jo Nye and Robert A. Nye, were appointed in 1994. Anita Guerrini and David A. Luft succeeded them in 2008. The Horning Visiting Scholar in the Humanities program was inaugurated in 2006 to allow a distinguished scholar to spend a week in residence at OSU and deliver a series of lectures as well as participate in other activities in and out of the classroom. Visiting scholars have included Liba Taub (University of Cambridge), John Beatty (University of British Columbia), Pamela O. Long (independent scholar), and Ken Albala (University of the Pacific).

The OSU Press Horning Visiting Scholars Publication Series, under the direction of the press's acquisitions editor, Mary Elizabeth Braun, publishes the public lectures delivered by the Horning Visiting Scholar: three volumes in the series have appeared, Liba Taub's *Aetna and the Moon* (2008), Pamela O. Long's *Artisan/Practitioners and the New Science* (2011), and Ken Albala's *Grow Food, Cook Food, Eat Food* (2013).

Robert Fox, Emeritus Professor of the History of Science at the University of Oxford, was Horning Visiting Scholar in May 2013. Educated at Oxford (BA Physics, DPhil History of Science), Professor Fox spent over twenty years at the University of Lancaster before returning to his alma mater in 1988, where he held the chair in History of Science until 2006. Professor Fox has specialized in the history of modern (post-1700) physical science, particularly in France. His numerous publications include *The Caloric Theory of Gases from Lavoisier to Regnault* (1971); *The Organization of Science and Technology in France, 1808–1914*, edited with George Weisz (1980); *The Culture of Science in France, 1700–1900* (1992); *Education, Technology, and Industrial Performance in Europe, 1850–1939*, edited with Anna Guagnini (1993); *Natural Dyestuffs and Industrial Culture in Europe, 1750–1880,* edited with Agustí Nieto-Galan (2000); *The Savant and the State: Science and Cultural Politics in Nineteenth-Century France* (2012); and, coedited with Jed Buchwald, *The Oxford Handbook of the History of Physics* (2013). Among his many honors, he is a Chevalier in the French Ordre des Arts et des Lettres, and, in 2015, the History of Science Society presented him with the Sarton Medal, its highest award for scholarly achievement.

In *Science without Frontiers: Cosmopolitanism and National Interests in the World of Learning, 1870–1940*, Professor Fox turns from science as a practice to science as a model for society. Scientific internationalism, fostered by international congresses and societies, flourished from the mid-nineteenth century until 1914. It promoted optimistic views of the future of mankind based on cooperation rather than competition.

Such disparate projects as the universal language Esperanto and Melvil Dewey's numerical system of library cataloguing sprang, Professor Fox argues, from the same internationalist (and pacifist) impulse. The sharing of knowledge across national boundaries was essential to scientific internationalism, and in chapter 1 Professor Fox focuses on the Institut international de bibliographie, founded in Brussels in 1895, as a center for this collaborative endeavor.

As related in chapter 2, the First World War, remarkably, did not entirely crush this effort. But its revival after the war under the auspices of the League of Nations contended with a new set of challenges as the usefulness of science and technology became manifest to European governments. Chapter 3 details the fate of cooperative scientific internationalism in Europe and the challenges posed to it by the rise of totalitarianism and the conflicting forces of nationalism and internationalism. Professor Fox looks at public expressions of scientific nationalism in museum exhibits and, most tellingly, in rival national pavilions at such celebrations of internationalism as the Paris International Exposition of 1937. The Second World War, it would seem, would have shattered internationalist ideals for good, but Professor Fox in an epilogue finds some grounds for optimism in the success over the years of UNESCO, and views Google Books and other electronic media as a new way to achieve the prewar internationalists' vision of universal access to knowledge.

Science without Frontiers offers a new way to think about science and culture and its relationship to politics amid the crises of the twentieth century. Professor Fox's deep knowledge of this era and its science enriches our ideas of the possibilities of a truly international science.

Anita Guerrini

PREFACE

Science without Frontiers has its roots in the lectures I was privileged to give as Horning Visiting Scholar at Oregon State University, Corvallis, in May 2013. Invited by Anita Guerrini, Horning Professor in the Humanities and Professor of History, I enjoyed a week of memorably stimulating discussions and generous hospitality that encouraged me to pursue the themes now developed in this book. Later in the year, the semester I spent in Philadelphia as Gordon Cain Distinguished Fellow in the Chemical Heritage Foundation's Beckman Center for the History of Chemistry was crucial in providing the scholarly environment that the early stages of the transition from the lecture format to publication required. In the Beckman Center, I benefited from discussion after a "Fellow in Focus" public lecture and a writing workshop led by Carin Berkowitz, as well as frequent discussions with other Beckman Center Fellows, especially Evan Hepler-Smth and Alex Csiszar, whose interests overlapped fruitfully with mine. Elsewhere, I have accumulated many debts to colleagues and students who have given encouragement and feedback in seminars and lectures at Yale, Oxford, the Université libre de Bruxelles, the Universitat Autònoma de Barcelona, and the Espace Fondation EDF in Paris, also at an international workshop, "Science en guerre et guerre des savants," hosted by the Académie royale de Belgique in November 2014, and at meetings of the Amitiés Françaises de Mons, the Brazilian Society for the History of Science, and the Italian Society for the History of Science.

It is a happy tradition of the Horning Professorships that Horning lectures are published by Oregon State University Press. From the submission of my manuscript to the Press's acquisitions editor, Mary Braun, through to the final stages of production, I have benefited from the friendliness and efficiency of a truly remarkable academic publisher. Only I can see how much *Science without Frontiers* owes to an eagle-eyed copyeditor Susan Campbell, the Press's editorial, design, and production manager Micki Reaman, and Tom Booth, associate director, who attended creatively to the book's illustrations and design. Among other friends in Corvallis, I am especially indebted to Anita Guerrini and David Luft, both of whom offered comments that led me to rethink important parts of my manuscript.

Science without Frontiers could not have been written without the collections that make historical research possible. Access to the Othmer Library of Chemical History at CHF in 2013 and on subsequent brief visits has been a special privilege, as have two visits to the Mundaneum Archives in Mons. The resources of the British Library, London Library, Bibliothèque nationale de France, and Library of Congress have all played their part as well, complementing the holdings of the Bodleian, my wonderfully rich "home" library in Oxford. Everywhere, librarians and archivists have been unfailingly helpful. But I owe a special debt to those who have provided images that appear in the book: Stéphanie Manfroid at the Mundaneum Archives; Cornelia Schörg and Erika Simoni in the Archive of the Technisches Museum, Vienna; Anna Krutsch of the Deutsches Museum Archive, Munich; Alicia Gómez-Navarro, Director of the Residencia de Estudiantes, Madrid; Matilde Amaturo and Paolo Di Marzio of the Museo Hendrik Christian Andersen, Rome; Francesco Camporini of the Fondazione Alessandro Volta, Como; and Valérie Lesauvage, documentalist at the Roger-Viollet photographic agency in Paris. The help I have received from these and other colleagues has exemplified the generosity and openness of the world of learning to which so many of those I treat in my book aspired.

Robert Fox
Oxford, September 2016

INTRODUCTION

Among the myriad forms of learned culture, science has traditionally had a special place. As a body of knowledge whose truths have been thought to depend on the use of reason, observation, and experiment, it has had connotations of a universality unbounded by national frontiers and cultural traditions. In the mid-seventeenth century, Henry Oldenburg saw *Philosophical Transactions*, the journal that he founded in 1665, as a depository for reports from "the ingenious," wherever they might be, and both his correspondence and the journal itself gave concrete expression to his conception of a limitless, undivided world of learning, united in the principle that knowledge should be accessible to any who cared to seek it.[1] In the eighteenth century, the same principle animated the cosmopolitanism of a Republic of Letters, in which correspondence flowed freely and Catherine the Great established figures of the international standing of Leonhard Euler and Peter Simon Pallas as long-term residents in St. Petersburg, close to the imperial court but with links to networks throughout the scientific world. In the nineteenth century, the tradition continued. Now, though, it did so in ways that reflected a new urgency for knowledge to be shared across the boundaries of language and local community. Especially from the midcentury, increased activity in transnational fields such as geodesy and meteorology, international projects for the mapping of the heavens (exemplified in the Astrographic Catalogue and Carte du Ciel, launched with collaborating observatories in more than a dozen countries in 1887), and the quest for agreements on nomenclature and the definition of physical units and standards focused minds everywhere on the value of collaborative endeavor and the mechanisms of communication and information retrieval that necessarily went with it.

A mark of the increasingly international character of science was the proliferation of congresses, almost unknown in the 1840s but part and parcel of science by the end of the century. Such events owed much to the unprecedented ease of travel and hence to the very advances in science and engineering to which many congresses were devoted. Expanding rail and maritime networks meant that scientists with the means and will to travel were no longer constrained by the limitations of what they could achieve through their home institutions and immediate personal contacts. The expansion of the periodical and monograph literature in science and the increased availability of translations of books and articles likewise extended intellectual horizons and gave new meaning to scientists' perception of themselves as members of a connected global community of the learned.

From slow beginnings in the midcentury, these trends gathered strength from the 1870s. They were facets of what Anne Rasmussen has identified as a tide of scientific internationalism that reached its peak in the quarter of a century or so before the First World War.[2] Her analysis, following the pioneering work of Brigitte Schroeder-Gudehus, points to the diverse forms that such internationalism could take and the incentives that drew governments, individuals, and national communities in the sciences to subscribe to its ideals. The consequences were manifest not only in congresses and their associated cooperative endeavors but also in the international exhibitions with which congresses were often associated.[3] Such exhibitions, beginning with the Great Exhibition of 1851 in London and catching on quickly across the Old and New Worlds, constituted a popular shopwindow for modern industrial society and the scientific and technological achievements underlying it. Their usual designation as "universal" conveyed more than the range of the exhibits on show; it also expressed the geographical compass of what Americans in particular referred to as "world" fairs.

In *Science without Frontiers* I argue that shared research goals and scientists' readiness to take advantage of the dramatically improved provision for communication across national and linguistic boundaries had much in common with contemporary internationalist movements extending far beyond the realms of science and technology. Driving many of these broader movements were hopes of a future world from which disharmony and misunderstanding might one day be eradicated, with the aid of ever more effective networks of communication. At the intergovernmental level, such new foundations as the International Telegraph Union (of 1865), the Universal Postal Union (created under the Treaty of Bern of 1874), and the Inter-Parliamentary Union (founded in 1889 to establish bridges between parliamentarians of all nationalities) were conceived in precisely this spirit. But equally striking transnational ventures owed their origins to groups of like-minded individuals acting with no or little governmental support. Pacifism, the incipient stirrings of international socialism, feminism, and anticolonialism, and the promotion of Esperanto and other constructed languages all grew in this way, fired by idealistic beliefs founded on the essential oneness of humanity and the need to overcome the obstacles that threatened that unity.

A political and cultural climate so conducive to the sharing of knowledge was fertile ground for the late-nineteenth-century innovations in cataloguing and bibliography that I discuss in chapter 1. These were famously exemplified in the success of Melvil Dewey's decimal system of classification, dating from 1876. But in ways more immediately pertinent to the themes of this book, the innovations inspired the creation of an international bibliographical institute (IIB, in accordance with its French title, Institut international de bibliographie) in Brussels in 1895. The IIB was at one level a practical response to the rising tide of publications of which scientists and others felt they should keep abreast. But it had aspirations that went beyond the challenge of late-nineteenth-century information overload, important and alarming though this was. The IIB's founders, Paul Otlet and Henri La Fontaine, were defined by their lifelong dedication to the higher and, for them, inseparable causes of peace and free access to information of all kinds and for all peoples. In 1910, they took their vision further in founding a Union of International Associations (or Union des associations internationales, UAI). The aim of the union was to complement the bridge-building efforts of governments by fostering links between academies and other societies and nongovernmental bodies. In this respect, the world they sought to promote was more properly "universal" than conventionally international.[4]

It is significant that both Otlet and La Fontaine were Belgian and that their bibliographical institute and the union had their headquarters in Brussels. While internationalist initiatives certainly took root elsewhere, in Switzerland and the Netherlands for example, Belgium was particularly receptive to the prewar internationalist spirit that Daniel Laqua and others have seen as more generally characteristic of the period. As Laqua argues, neutrality, guaranteed by the Treaty of London in 1839, gave the country a special status. Belgium was also a nation whose aspirations (since its creation in 1830) had been bound up with its geographical position as (in Louis Piérard's words) "a corridor for the transit of values," in which ideas were traded between the Latin and Germanic worlds.[5] A location and cultural traditions of such particularity had consequences. The area that we know as Belgium had suffered for centuries as a target for foreign domination. It was a dismal distinction, but one that made Antwerp a natural setting for the Universal Peace Congress of 1894 and Belgians a conspicuously peace-loving people.[6]

Laqua's analysis bears centrally on themes in the present book in its demonstration of the inextricability of the forces of nationalism and internationalism. He shows convincingly how the Belgian government's commitment to internationalist initiatives (including the bibliographical institute in Brussels) before 1914 went hand in hand with a powerful sense of national identity. Internationalism, as Laqua puts it, formed part of the country's "national discourse" and was an essential prop of Belgium's standing as a nation.[7] Other recent studies have pointed no less forcefully to patterns of coexistence between internationalist sentiment and keen sensitivities to nationality in countries and in periods beyond the half century from 1880 that Laqua identifies as Belgium's "age of internationalism."[8] These broader perspectives are explored in a landmark collection of essays, *The Mechanics of Internationalism* (from the 1840s to the First World War), in studies by Madeleine Herren and Sacha Zala, which take a comparative perspective embracing Switzerland and the United States as well as Belgium (through to 1950), and in Glenda Sluga's book on internationalism from the turn of the twentieth century to our post–9/11 world.[9] As Martin Geyer and Johannes Paulmann make clear in their introduction to *The Mechanics of Internationalism*, internationalism and nationalism were not necessarily polar opposites, at least in the period they treat, to 1914. While, in some contexts, they did exist in opposition to each other, they could also work together in the national interest. In Geyer and Paulmann's words, internationalist ideals should not be seen therefore as just a "weak alternative to the power politics of the nation-states" and hence as unsubstantiated expressions of optimism to which historians need to pay little attention.[10]

In trying to integrate science and technology in these bigger pictures, we should be similarly cautious. Museums, international exhibitions, and congresses, for example, always had their competitive, "Olympic" dimension, well described by Svante Lindqvist and Geert Somsen.[11] Yet the intrinsically transnational character of scientific knowledge meant that national as well as personal rivalries tended to be expressed discreetly, even hidden from the public record completely. That at least was the case before 1914. As I argue in chapter 2, it took the First World War, a conflict of unprecedented brutality and compass, to shatter the prevailing tone of shared endeavor. When it happened, the change came about with remarkable rapidity. Barely two months into the war, on October 4, 1914, ninety-three German intellectuals, including many eminent scientists, published a manifesto proclaiming the duplicity of the Allies and the rightness of the cause for which the Central powers were fighting. And thereafter insults and justificatory prose flowed freely, in both directions. In what the contemporary German politician Hermann Kellermann presented as an intellectuals' war between hitherto otherworldly academic "spirits," many who had once prided themselves on being citizens of a world without frontiers now pledged allegiance to their countries.[12] As part of that change, belittling enemy scientists who had so recently been collaborators, even personal friends, suddenly became an act of patriotism.[13]

The readiness of the scientific communities of the belligerent countries to engage in such reciprocal disparagement invites the sceptical conclusion that the internationalist rhetoric of prewar science had been a veneer that concealed the reality of a world riven by intense competitiveness. Perhaps scientific internationalism all along had been merely the stuff of dreams. In this book, I resist that stark interpretation, which I believe owes too much to the distorting prism of two world wars and a protracted cold war through which we have to view this segment of the past. Certainly, in exchanges of the kind that led to the naming of a new unit, species, or organic compound, national interests could never be entirely overridden. Yet the dominant image, of science as a pursuit capable of transcending nationality, has to be seen as more than an expendable strategic charade. It reflected and informed the day-to-day realities of relations between national communities and individuals in the sciences. And it helped make science a model for a world order in which shared understanding across all fields of knowledge, and not just the sciences, offered the best prospect of peace.

The Brussels-based initiatives of Otlet and La Fontaine were an expression of precisely that aspiration. And they were not alone. Indeed, in their global aims, even these bold ventures could not match the grand architectural project of the Norwegian-American artist and visionary Hendrik Christian Andersen for a city that he presented as a "world centre of communication." Andersen's city had a dreamlike quality that exposed it to scepticism, in some cases ridicule, when he published his plan in 1913. Many, though, took it seriously, at least in its overall conception, if not in its plethora of extravagant details. We should certainly not dismiss it as an empty idealist fantasy. It drew on and helped inform a significant current of contemporary opinion, albeit one that suffered a crippling blow when the First World War began soon afterward.

In casting the war and its legacy of recrimination as a "watershed" as I do in chapter 2, I stress the consequences of the carnage and mistrust of 1914–1918 for prewar conceptions of a seamless community of the learned. Fritz Haber's engagement in research on poison gases encapsulated the readiness of scientists to pursue their nations' interests in what contemporaries recognized as an uncomfortable clash between the ideals of scientific openness and patriotic duty. Haber, of course, was by no means alone. Scientists on both sides of the conflict engaged in whatever branch of their country's war effort they could best serve, and in the process turned from openness to secrecy as the norm of science. After the Armistice, prewar champions of the free exchange of knowledge did what they could to refurbish their vision. In 1919, Andersen published a supplement to his plan, and in the late 1920s Le Corbusier promoted his own plan for a "cité mondiale," a world city similar to Andersen's, intended for construction (though, like Andersen's, never built) near Geneva. Otlet and La Fontaine too were quick to resurrect their bibliographical institute and the Union des associations internationales, and international congresses and exhibitions soon got under way again. The world, though, had changed, irretrievably. Quite apart from the barely dimmed enmity between the former belligerents, the fluid voluntarism, backed by intense personal commitment, that had done so much to sustain the internationalist cause before 1914 was no longer enough. This change weighed particularly heavily on Otlet, as he saw core functions of the union being gradually taken over from 1922 by the International Committee on Intellectual Co-operation, the better-financed and better-organized cultural arm of the League of Nations.

Chapter 2 extends discussion of the watershed of the First World War into the early 1920s, to bring out the continuing effect of the animosities engendered by a war that in important respects did not end with the Armistice or even the Paris Peace Conference of 1919. Institutionally, a major obstacle to any prospect of a rapid return to the prewar ideal of an undivided community in science was the International Research Council, a body that the Allied powers created in 1919 to coordinate the activities of newly established unions for the various sciences. The council's stated mission was cast in terms of cooperation between nations. But the internationalism that resulted was profoundly colored by memories of the war. Membership was limited to the wartime allies and, subject to their approval, such neutral nations as might wish to apply. The overriding principle was that the former Central powers should be rigorously excluded, both from the council and from the disciplinary unions. Such unforgiving obduracy, in particular with regard to a scientifically renascent Germany, created profound tensions. To the gathering concern of the member nations, to say nothing of those that were excluded, the council had blatantly subordinated the interests of science to the vindictiveness of victor's justice.

The exclusionist stand of the International Research Council, as also of the International Committee on Intellectual Co-operation (with the resolutely internationalist Einstein as its only German member), was part of what I see as a "national turn" that gathered strength in science as in other areas of intellectual life between the two world wars. Underlying it was the rise of science and technology in governmental priorities everywhere. What occurred was not wholly new; support for scientific and technological research had been an element in the budgets of most advanced countries since the mid-nineteenth century. But the war accelerated a hitherto rather patchy recognition of science as a major governmental responsibility and a source of prestige and economic and military strength. The benefits that accrued from the process through the 1920s and even in the difficult years that followed the crash of 1929 are evident: most obviously in the United States, though in varying degrees elsewhere too, new laboratories and enhanced funding for scientific research and teaching went hand in hand with a generally improved status for scientists. As conduits of new governmental favor toward science, the Department of Scientific and Industrial Research (a wartime creation in Britain) and the Consiglio Nazionale delle Ricerche (founded in Italy in 1923) were just two consequences of invigorated national science policies that had their counterparts in most countries.

As events were to show, however, the heightened profile of science in the affairs of state came with its dangers too, especially in nations under totalitarian rule. The constraints on scientific freedom in the Soviet Union in the years of collectivization and Stalinist purges beginning in 1929 and the enforced flight of Jewish scientists from Nazi Germany after 1933 and then Fascist Italy from 1938 were tragic reminders of how easily scientific priorities could fall victim to politically charged national expediency. In a different register but no less revealingly, national museums and international exhibitions assumed new roles as instruments of propaganda. In chapter 3, I illustrate the point with reference to the international exhibition of 1937 in Paris and the one planned (though never executed) for 1942 in Rome, both of which offered settings in which totalitarian regimes of right and left could display their particular conceptions of modernity and progress.[14] But other exhibitions too—the "century of progress" exhibitions of the 1930s in the United States, for example—show how readily events cast as "universal" (in the official designation determined at a diplomatic conference on international exhibitions in 1928) lent themselves to the pursuit of political ends. In the American case, the original Century of Progress exhibition of 1933 in Chicago, the Golden Gate International Exposition of 1939 in San Francisco, and the New York World's Fair in 1939–1940 (among the most prominent examples) had an important role as declarations of US hopes of prosperity despite the effects of the Great Depression.[15]

So was the interwar story a straightforward one, of the gradual retreat of a congeries of ideals and practices that contemporaries and historians have variously and often loosely described as international, transnational, cosmopolitan, or universal? Was retreat the inevitable consequence of a devastating Great War that had relegated prewar internationalism to the realms of naïve unrealizable fantasy? Certainly the ideals were hard-pressed to survive in anything resembling the form they had had before 1914. But they never disappeared. When a renewed world conflict came in 1939, they were sorely tried once again, and once again they weathered the storm. From 1946 they found expression in the newly created UNESCO, a successor in many respects to the International Committee on Intellectual Co-operation. And they have remained fixtures in the rhetoric of science. International congresses and unions continue to draw on these ideals, as do institutions of quite recent years, such as the Academia Europaea and the European Academy of Sciences, both committed to the promotion of exchange and interaction across national frontiers.[16] Our burgeoning age of open access to the world's scientific literature—and techniques of communication and an ease of travel of which Otlet, La Fontaine, and Andersen could only have dreamed—continue to open new vistas. In the twenty-first century, we may never recreate a "Scientific International" imbued with the optimism that Anne Rasmussen identifies as characteristic of the years immediately before the First World War.[17] But the core aspirations of that more hopeful age endure and still have power to fashion the fabric of international science in our own time.

CHAPTER ONE

Knowledge, the Cement of Nations

Through the nineteenth century, the ordering of knowledge was seen as a matter of acute and growing concern. For those who possessed knowledge and those who sought access to it, the most obvious problem was its sheer quantity. Knowledge, quite simply, showed every sign of spiralling out of control and of adding a new dimension to the age-old question regarding the amount of what an educated person might be expected to read and retain. The result was a swell of rueful reflection on the passing of a world in which, as some believed, it had once been possible for an individual to claim mastery of everything there was to know. The reflection, of course, rested on a fantasy about an age that had never really existed. But the dream persisted, from Athanasius Kircher in the seventeenth century to Thomas Young at the beginning of the nineteenth century, and it only faded thereafter as advancing specialization and the sheer amount of material to be read inexorably put the goal beyond the reach of any one person.[1]

By the mid-nineteenth century, even the omniscience that was commonly ascribed to the prodigiously versatile Cambridge philosopher and scientist William Whewell went hand in hand with a reputation for superficiality. As the wit and critic Sydney Smith put it, while science was Whewell's "forte," omniscience was his "foible": despite his unquestionably fine mind, even Whewell could not master all the disciplines in which he worked in the course of his long and intellectually busy career.[2] A generation later, the intellectual life of French organic chemist Marcellin Berthelot conveyed the last dying embers of the universalist dream. Obituaries and an early biographer portrayed him as the last savant who succeeded in embracing the whole realm of human knowledge.[3] As a polymath who wrote extensively on educational policy, philosophy, and the history of alchemy, Berthelot certainly knew many things. But the command to which he aspired had been recognized as unrealistic since long before his death in 1907. The career of a younger contemporary, Henri Poincaré, equally illustrates the point. Like Berthelot, Poincaré took a virtually boundless view of the field of knowledge, with interests embracing philosophy as well as the whole realm of mathematics. But when he died, five years after Berthelot, the capaciousness of Poincaré's intellectual engagement was seen, like Berthelot's, as marking the end of a by now unrealizable dream.[4]

It is easy to see why anxiety about the management and retrieval of knowledge should have come to be felt with particular keenness in the later nineteenth century. Even for those who limited their reading to the sciences, the escalating proliferation of books and journals engendered a sense of what we would now call information overload. The increasingly international character of science, promoted by congresses and improved channels of communication of every kind, served only to aggravate the concern. In simply keeping abreast of the essential literature in their own field, scientists faced a formidable challenge that forced a rethinking of older, more circumscribed habits of reading and citation.

In response to the growing concern, new departures in librarianship, bibliography, and means of access to sources assumed the status of practical necessities. But these seemingly workaday innovations quickly became invested with the higher ideal of a world in which knowledge of all kinds would circulate without regard for nationality or linguistic tradition. In an age striving for international harmony but beset with rivalries of the kind that drove European powers to pursue their competing colonial interests in Africa from the 1880s, the free exchange of knowledge came to be seen as an essential prerequisite for understanding between nations and cultures and hence as a bulwark against conflict. As such, it was a cause to which some of the leading champions of internationalist movements, most of them with a strong pacifist streak, gave their enthusiastic backing. It was in this context that concerns rooted in the mounting challenges of information retrieval became inextricably involved in anxieties of a very different order, for world peace.

Transcending boundaries

As Ann Blair has shown, concern at there being "too much to know" has been with us since antiquity.[5] Socrates is said to have encapsulated a sign of the times in his concern at a follower's recourse to note-taking rather than the well-tried but already increasingly demanding exercise of memory.[6] But the invention of the printing press in the fifteenth century unquestionably exacerbated the problem. When Ben Jonson rather circuitously endorsed the Senecan tag "Patet omnibus veritas" in his seventeenth-century commonplace book, he probably saw the statement as self-evident.[7] In principle, truth was indeed open to all. Yet it was only fully open to those who knew how to get at it, and by Jonson's time the sheer volume of what was now overwhelmingly printed material was already making access to truth difficult.

Statistics about the production of books are striking. The number of incunabula (books published in the first half century or so of printing, to 1501) did not exceed thirty thousand, many of these being new editions and reissues of classical and other texts inherited from the medieval tradition. But by 1600, the thirty thousand titles had become three hundred thousand, and the number was growing rapidly, along with the ever greater diversity of subjects treated.[8] The guardians of libraries, public and private, responded with an armory of traditional techniques. Two new foundations—the university library in Leiden, inaugurated in 1575, and the new Bodleian Library of 1602 in Oxford—were typical in adopting the well-tried medieval model of books classified by subject and housed in clearly designated cases. It was a sign of the times, however, that in 1595 the Leiden library supplemented its provision for readers with a printed catalogue, the first printed library catalogue that we know of.[9] The Bodleian and other major libraries soon followed suit, in an attempt to make their holdings known across the world of learning.[10]

By comparison with the core challenge of the number of books to be housed and catalogued, another potential impediment to Jonson's ideals of openness and accessibility, that of language, presented as yet a minor difficulty. But the incipient decline of publishing in Latin was a sign of difficulties to come: the proportion of the accumulated body of books published in Latin had fallen from 70 percent in 1500 to 50 percent a century later.[11] Even so, in 1600, most scholars still coped well enough by switching between the linguistic registers of Latin and a vernacular. The English polymath and author of the first description of North America in English, Thomas Harriot, was typical. In choosing English for his *Briefe and True Report of the New Found Land of Virginia* of 1588, he was probably adjusting to the expectations of his patron, Sir Walter Raleigh, who had financed the voyage of 1585–1586 that took Harriot to the outer banks of what is now North Carolina. Yet when the book was reissued in 1590, Harriot turned to Latin in writing commentaries for the illustrations by John White that now accompanied the text. His original, elegant Latin was then translated into English, French, and German for the editions in those languages published by the Flemish engraver and publisher Theodor de Bry in 1590.[12]

Examples of linguistic versatility of this kind remained common enough a century later. Newton, for example, was equally at home, at least for scientific purposes, in Latin or English: of his two great works, the *Principia* appeared in Latin (1687), the *Opticks* in English (1704). But by Newton's day the decline of Latin as a shared scholarly language was beginning to gather pace, though still slowly. It is a measure of Latin's resilience that in the later seventeenth and early eighteenth centuries, the Royal Society conducted almost a quarter of its correspondence in Latin, and that 14 percent of the surviving copies of oral presentations and papers submitted between the 1660s and 1730 are also in Latin.[13] As Henry Oldenburg recognized from the time he launched *Philosophical Transactions* in 1665, if the contributions he published in the journal were to be known abroad, a translation into Latin remained the best vehicle, not that his plan for an authorized Latin edition ever materialized.[14] Of other possible languages, none had the near universal quality of Latin.

It is significant that in the early Royal Society, even work in the best-known foreign language, French, was routinely translated into the society's working language, which was English from the start.[15]

Through the eighteenth century, especially in the sciences, Latin continued its retreat, as academies and other learned societies displayed their modernity by choosing to publish in whatever they deemed the most appropriate vernacular.[16] The chosen language might be a foreign one: in Berlin from 1746, for example, the reconstituted Académie royale des sciences et belles-lettres abandoned Latin and followed Frederick the Great's preference in opting for French rather than German (which Frederick viewed with contempt). But generally academies in the eighteenth century, both new and old, favored the languages of local speech: French in the Paris Académie des sciences and the new Académie impériale et royale des sciences et belles-lettres in Brussels (founded in 1772, in the time of Austrian domination), Swedish in the Kungliga Svenska Vetenskapsakademien in Stockholm (1739), Spanish in the Real Academia de Ciencias y Artes de Barcelona (1764), Portuguese in the Academia Real das Ciencias de Lisboa (1779), Italian (increasingly, albeit with an admixture of French and some residual Latin) in the Reale Accademia delle Scienze di Torino (1783), and English in the newly formed American Philosophical Society, where in 1743 the use of Latin would have been hard to reconcile with the society's modern-minded spirit. By the time the reformed St. Petersburg Academy signalled a new departure by changing its official name from Academia scientiarum imperialis Petropolitana to Académie des sciences in 1803, its continued use of Latin for over a third of the papers presented at its meetings had made it an outlier in a learned world that by now had effectively abandoned its once-common vehicle of communication.[17]

The process of linguistic fragmentation went hand in hand with the continuing challenge of the accelerating production of books. The three hundred thousand titles that had been published by 1600 had grown to an accumulated three million by 1800, a figure that had reached nine million by 1900, then twelve million in the 1930s, on its way to the 130 million or so volumes in our own day.[18] Periodicals followed a similar trajectory. More than half a century ago, in a classic study of what he called the "diseases of science," Derek Price identified about ten journals with broadly scientific interests in the mid-eighteenth century and mapped a pattern of exponential growth, in both their number and their contents, that went on to outstrip the capacity of even assiduous scientists to keep up with publications in their discipline.[19] Relentlessly driving the trend in the nineteenth-century periodical literature, as Alex Csiszar has argued, were changing patterns of career-making and reputation-building.[20] As successful careers and status came to be constructed increasingly around a record of publication in specialized journals, scientists turned away from correspondence and face-to-face contact as the primary ways of marking priority. The Scottish physicist William Thomson was someone who lived this transition. In a sixty-year career that began in the 1840s, when the published scientific paper was beginning to establish itself as the standard means of communicating research, Thomson published almost seven hundred papers, an average of one a month. It was a brisk pace but not a surprising one for a scientist so committed to research over such a long period.[21]

Amid this sea of print, what did it mean by the end of the nineteenth century to say, as Jonson had done almost three hundred years before, that truth lay open to all? Even if the truth in question was limited to the sciences, or even to a single science, the principle that knowledge should be readily accessible to anyone with the capacity and will to seek it out had become ever harder to realize. One cause of this lay in the expanding geographical scope of much of science since the later eighteenth century. Such nascent fields as geodesy and meteorology simply did not make sense if they were pursued in a purely national context, and it is no coincidence that practitioners of these two sciences were among the pioneers of systematic initiatives in transnational collaboration: the International Meteorological Organization (founded in 1873) and the International Geodetic Association (founded in 1886, on the basis of earlier initiatives led by Prussia) were among the first discipline-based organizations to work systematically across national boundaries. Established disciplines too were set on increasingly international trajectories. Their needs were essentially practical, for systems of nomenclature and other conventions that would enhance precision and clarity. With regard to nomenclature, botany (with Linnaeus) and chemistry (with Lavoisier and his circle) were two sciences that had already laid enduring foundations during the eighteenth century. But after 1800, as the pace of fieldwork and experimenting increased, international agreement in the naming of specimens and the reporting of observations became a necessity in all the observational sciences.

In physics and engineering, the pressures were somewhat different. Here the nineteenth-century trend toward standardization and precise measurement enhanced the demand for more accurate determinations of the metric units of length and weight and, as telegraphic communication advanced from the 1860s, for exact definitions of the ohm and other units of electromagnetism.[22] In a technology founded on the detection of small currents and subject to the vagaries of the electrical properties of the cables, telegraphy helped to propel quantitative precision, exemplified in the quest for standard resistance coils of extreme accuracy, to the forefront of physicists' and engineers' concerns. With these incentives, agreements thrashed out at congresses and on international committees assumed an importance that extended far beyond curiosity-driven science.

Scientists everywhere welcomed the benefits that came with the intensification of cooperation and exchange across national boundaries. Those whose research lent itself to diffusion to audiences in other countries were happy to see their work translated and published abroad. J. C. Poggendorff's *Annalen der Physik* (later the *Annalen der Physik und Chemie*) was a pioneer among German journals that regularly published translated articles in full or abbreviated form from its foundation in 1799. But the practice became conspicuously more common from the midcentury. In the English language, the five volumes of translated articles that Richard Taylor edited and published between 1837 and 1852 were a notable new venture.[23] And in France, Adolphe Wurtz and Emile Verdet were just two of a younger generation of French scientists who undertook translations, mainly of German and British articles, for publication in the *Annales de chimie et de physique* from the 1850s. In a journal that had hitherto been largely devoted to the products of French science, the systematization of the coverage of work published abroad marked a significant turn.

Books too became an important vehicle for the circulation of science as an international commodity. Among the most ambitious and enduring initiatives in the field was the collection launched in its English-language form as the International Scientific Series in the 1870s. The idea of the series grew from a proposal at a meeting of the British Association for the Advancement of Science in Edinburgh in 1871 for an agreement that would promote the simultaneous publication of key scientific works in different languages. Within barely a year, the project was launched, and it continued until 1919, in a partnership between the publishing houses of Kegan Paul in Britain, Germer-Baillière (later Félix Alcan) in France, Appleton in New York, Brockhaus in Germany, and Fratelli Dumolard in Italy. The collaboration was never perfect: by no means all books appeared in all the languages of the agreement, and the lists in some languages were small. A Russian venture involving the publishing house of Znanie in St. Petersburg, for example, barely got off the ground. But the English and French series in particular proved so successful that by 1900 Félix Alcan could boast a list of ninety-three French titles in his Bibliothèque scientifique internationale, with an accumulated total of 122 by 1919.[24]

It was with a similar ambition to strike out beyond their various national traditions that nineteenth-century scientists embraced the genre of the international congress. The practice of meeting at congresses was not new. But congresses in the first half of the century were largely national affairs: at gatherings of the Gesellschaft Deutscher Naturforscher und Ärzte (founded in Leipzig in 1822), the British Association for the Advancement of Science (annually from 1831), the French Congrès scientifiques (from 1833), and the Congressi degli scienziati italiani (rather irregularly from 1839) visitors from abroad were often present, but in small numbers. Truly international congresses were a phenomenon of the later nineteenth century. From a small handful in the 1840s and 1850s, the number of these had grown dramatically by the last decade before the First World War (see figure 1).[25] In a total of some twenty-seven hundred such events over these sixty years, the range of subjects was vast. Political, commercial, and moral themes were all strongly represented. But so too was a broad spectrum of scientific and scholarly disciplines.

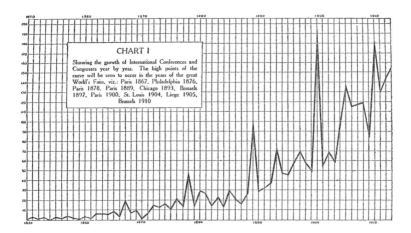

1. Number of international conferences and congresses by year, 1850–1913. The peaks coincide with major international exhibitions. Those in 1900 and 1910, the years of the exhibitions in Paris and Brussels, are particularly marked. From John Culbert Faries, *The Rise of Internationalism* (New York: W. D. Gray, 1915), 74. In compiling this graph, for his doctoral thesis in political science at Columbia University, Faries used data in the *Annuaire de la vie internationale for 1910–1911* and the calendars of international meetings regularly published from 1912 in *La Vie internationale: Revue mensuelle des idées, des faits et des organismes internationaux.*

By the late nineteenth century and on through to the First World War, there were some forty congresses a year in science, medicine, or technology.[26] For scientists everywhere they were essential meeting places. When they were held in association with a major universal exhibition, as often happened, they assumed special visibility and significance for the discipline or activity concerned. The Exposition Universelle in Paris in 1900 was such an occasion. No fewer than 127 congresses met in Paris in that year. Of these, well over a third were on science-related themes, and many attracted several hundred delegates.[27] Such gatherings provided opportunities not only for the general strengthening of bonds between individuals and communities but also for the promotion of organized collaborative research.

In this and other ways, the popularity of congresses was both an indicator and a motor of internationalist trends in intellectual life. It was at one such congress, the Astrophotographic Congress of 1887 in Paris, that the most ambitious of all multinational research programmes, the Carte du ciel project, was launched.[28] In the course of ten days of discussions during the congress, some fifty astronomers from twenty countries, presided over by the director of the Paris Observatory, Amédée Mouchez, launched a plan for the creation of a definitive map of the heavens. Stars as faint as 11 magnitude, in both the southern and the northern hemispheres, were to be photographed and recorded in a multivolume Astrographic Catalogue, eventually completed in the 1960s. In such work, as in the associated plan for a second set of photographic plates recording stars down to 14 magnitude (constituting the true Carte du ciel project), easy cooperation and uniformity in the innovative photographic techniques and methods of recording were essential. The project began strongly, with twenty-two observatories across the world (though none in the United States or Russia) taking responsibility for particular declination zones of the sky. And for some years it stood as a model of what a well-organized multinational venture might achieve. If the project faltered, as it did after the First World War, the cause lay as much in the problems of the postwar economy and the immensity of the enterprise as in the weakening of the international spirit that had provided its initial impetus, although all these factors were at work.[29]

Equally international in its conception, though in the long term more successful, was the Bureau international des poids et mesures (BIPM). The bureau owed its origins to the international Convention on the Metre, signed by seventeen countries in 1875, following the lengthy Diplomatic Conference convened in Paris by the French government.[30] As part of the convention, it was agreed to establish the bureau as an intergovernmental center with offices and state-of-the art laboratories in the Château de Breteuil at Sèvres on the western outskirts of Paris. There, from 1878, the bureau supervised the prototypes and ultimate definitions of the metric system, especially the prototypes of the kilogram and the meter, through research, the calibration of metric standards, and the promotion of meetings, most importantly the three General Conferences on Weights and Measures in 1889, 1895, and 1901. Everything about the institution was international. Its first director was the Italian physicist Gilbert Govi, and Swiss, Norwegian, and French successors followed through to 1915. By its statutes, practices, and international panel of collaborators and staff, the BIPM won virtually uncontested recognition as the arbiter not only in matters concerning the metric system but also, in due course, in areas of physics as diverse as thermometry, electrical units, and photometry.

The collaborative spirit exemplified in work at the Château de Breteuil did not betoken a complete disregard for national interests. The decision to locate the BIPM in France, for example, was unquestionably a source of French national pride; coming as it did so soon after the country's defeat in the Franco-Prussian War, it amply compensated for the inevitable relaxing of the unchallenged control of the metric system that France had exercised since the French Revolution. Likewise in other fields, it mattered to the organizers of an international congress or exhibition that their country had been chosen as a venue, as it also mattered to national delegations that their favored principles of nomenclature or system of physical units and standards should be adopted in preference to those of their rivals.

Even the International Association of Academies (IAA), established in 1899 as a vehicle for bringing together the world's leading academies and with cooperation across national frontiers as a main function, was marked by similar ambiguities. From an initial group of nine academies (all European, except for the National Academy of Sciences in Washington), it grew rapidly to a membership of twenty-four academies and other societies in sixteen countries by 1914. Along the way, its mission to promote cooperation without regard to nationality had its successes: an edition of the complete works of Leibniz, collaboration in solar research, and attempts to coordinate meteorological stations throughout the world were significant achievements with international scope.[31] But the IAA never entirely threw off suspicions that arose from its origins in an existing loose "cartel" of German academies. Although in the sciences the Paris Academy of Science and the Royal Society supported it from the start, they seized every opportunity of tempering the German tone that marked the IAA's early years by bolstering membership and activities beyond German-speaking Europe.

It might have been expected that the known firmness of Alfred Nobel's commitment to the internationalist cause would protect the Nobel Prizes from such tensions. But within Sweden divergent attitudes were apparent in discussions of how far national interests should be sacrificed in the name of the ideals imposed by Nobel's will shortly before his death in 1896. Chauvinist elements in Swedish society, including the king, Oscar II, criticized the prizes for diverting funds to foreign recipients at a time when the country's own scientists and scholars lacked the support they needed if they were to be properly recognized beyond Scandinavia and so assume the place to which they aspired in the world of learning.[32] And even among those who believed that the prizes should be awarded to recipients irrespective of their nationality (as Nobel intended), some argued that the Nobel committees should discount international trends and focus rather on aspects of science close to traditional Swedish strengths. The king's preference on this count was clear. His ideal Nordic scientist, and ideal prize winner, was Adolf Erik, Baron Nordenskiöld, a polar explorer and geologist known for his heroic quest to discover the long-sought Northeast Passage, which he described in a five-volume account of his voyage on the *Vega*.[33] Nordenskiöld's world was that of untamed nature rather than an academic research laboratory, and in that choice of workplace he embodied, for the king (who had sponsored some of his explorations) as for many conservatively inclined Swedes, truly Nordic values.

It is a mark of these divergent perspectives that national interests were never wholly eliminated from the selection process, despite the powerful guiding hand of the influential Swedish physical chemist Svante Arrhenius, who was conspicuously outward-looking in his outlook and a frequent visitor to Britain and the United States.[34] Arrhenius was almost certainly among those who were disappointed by the decision that the 1912 prize for physics should go to Gustaf Dalén for his "flasher," a regulator that controlled the flow of acetylene gas in lighthouses so that the light could be turned off automatically and fuel saved in daytime. Although the device lacked nothing in ingenuity, and its importance for shipping in Scandinavian waters was beyond question, it was hardly worthy of a prize at a time when the discipline of physics harbored so many distinguished candidates. The award to Dalén, made by the Swedish Academy on the basis of one rather perfunctory nomination and in opposition to the advice of its physics committee, was an early example of what Geert Somsen has identified as "Olympic internationalism," internationalism with a nationalistically tinged competitive edge in the manner of the Olympic Games.[35] Happily for the prizes' reputation, the balance was quickly rectified when mainstream internationalism reasserted itself in the awards that followed, to the Dutch physicist Kammerlingh Onnes in 1913 and to the German Max von Laue in 1914.

Bibliography and world peace

The gathering proliferation of international ventures in science from the mid-nineteenth century owed much to a recognition of the practical value of large-scale projects and the need for a controlled standardization of units, nomenclature, and the procedures of chemical analysis and precise physical measurement, were collaboration and the exchange of information to be effective. But more visionary aspirations—fired, in large measure, by a belief in the power of shared knowledge to promote the cause of peace—were at work as well. It is no coincidence that this facet of internationalism had its most resolute champions in Belgium, a country whose independence and neutrality, guaranteed in perpetuity by the European powers that signed the London Treaty of 1839, had long made it a natural home of movements dedicated to building bridges between nations.[36] And it was there, in Brussels, that Paul Otlet and Henri La Fontaine, two of the most prominent internationalists of the next four decades, began to make their mark soon after they first met in 1891.[37]

Until their causes began to fade in the 1930s, Otlet and La Fontaine, both lawyers by training and (at least in La Fontaine's case) profession, worked tirelessly to promote their dream of a "universe of information" (W. Boyd Rayward's expressive phrase) that would be made accessible to all through an exhaustive program of cataloguing and classification embracing knowledge in all forms and every language. Of the two, Otlet (age twenty-three at the time he met La Fontaine) was the more retiring, a meticulous, indefatigable doer, given to depression in his early life but blessed with a capacity for organization and meticulous attention to detail. La Fontaine, fourteen years his senior and already a noted jurist by the time his collaboration with Otlet began, was more urbane and is now best known as a well-connected public figure who won the Nobel Peace Prize in 1913, after a quarter of a century of engagement in the International Peace Bureau and related bodies. At once a patriot and an internationalist, La Fontaine went on to represent Belgium as a member of his country's delegation, first to the Paris Peace Conference of 1919 and then to the League of Nations General Assembly.

The overriding ideal of world peace lent everything that Otlet and La Fontaine did a dimension that went far beyond the immediate matters of call numbers and bibliographical accuracy. Yet the technical work was essential. A crucial first step was the creation of an Institut international de bibliographie (IIB), which they established in Brussels in 1895 following an inaugural international conference on bibliography, also in Brussels and also organized by them under the presidency of another noted campaigner for peace, Chevalier Edouard Descamps.[38] The IIB was a breathtakingly ambitious undertaking. With 3" x 5" cards, stored in labelled drawer cabinets, as its core recording device, the system stood at the cutting edge of contemporary library science (figure 2). Its call numbers were based on the system that Melvil Dewey had developed from the late 1870s, first at Amherst College and then in the library of Columbia University and the New York State Library. But the IIB's call numbers were longer and gave more information than those of the Dewey, allowing for greater exactness in recording not only authors, titles, and dates of publication but also the subject, language, and format of the item being catalogued.

2. The Institut international de bibliographie, Brussels, c. 1910. At its peak shortly before the First World War, the institute employed a staff of over twenty, with the task of creating a universal world bibliography organized in accordance with the IIB's Universal Decimal Classification. Courtesy of the Mundaneum, Mons.

The extended call numbers of what was codified in 1905 as the Universal Decimal Classification (UDC) offered more than pedantic precision. Since the IIB set out to catalogue images, films, and sound recordings as well as the usual printed items, the new call numbers were indispensable. Three-dimensional objects too were treated as sources of knowledge, with attendant challenges to cataloguers. When in 1910 Otlet and La Fontaine began assembling materials for an international museum (many salvaged from the Brussels Exposition Universelle of that year), they classified the collection in accordance with the UDC. And the same system guided the arrangement of the exhibits when, soon afterward, the government gave permission for them to be displayed in one of the capital's most prestigious locations, the Palais du cinquantenaire, constructed in 1880 to mark the fiftieth anniversary of the birth of the Belgian nation (figure 3).[39] The installation of the museum in this monumental exhibition space was a triumph for Otlet and La Fontaine. And a further accolade followed when the philanthropist Andrew Carnegie visited in 1913.[40] Carnegie, it seems, was charmed: as he wrote afterward, "I never enjoyed a visit more . . . I am astonished at what I found."[41] By now the museum occupied sixteen rooms and received almost thirteen thousand visitors a year.[42]

3. The telegraphy section, Musée international, Brussels. Paul Otlet and Henri La Fontaine conceived the museum as an extension of their plan for a universal bibliography. Installed in the Palais du cinquantenaire, the collection was built on a core of exhibits acquired after the highly successful Exposition Universelle of 1910 in Brussels. The dense, didactic style of presentation was typical of most of the museum's displays. Courtesy of the Mundaneum, Mons.

As a rationally planned presentation of material culture on a world scale, the museum was a natural extension of Otlet and La Fontaine's initial plan. Nevertheless, the core work of the IIB remained bibliographical, and it was from there, as head of the institute (though with the title of secretary-general), that Otlet set out, in the words of Alex Wright's biography of him, to catalogue the world.[43] The daunting project got off to a brisk start. A recurrent governmental grant delivered through the Belgian Ministry of the Interior and Public Instruction helped the IIB's staff of twenty-three process some two thousand cards a day at its peak. The number of cards reached half a million by the end of its first year of operation and eleven million by 1914. The figures convey the ambition of the project. Its goal was a "universal" resource that would surpass existing national bibliographies not only in the range of materials covered but also in the ease of access that it promised.[44] Despite the IIB's Belgian roots, it was essential to its mission that the resource should be shared: to this end, satellite depositories for the cards were planned for Washington, Paris, Rio de Janeiro, and other cities across the world. From these centers, as from Brussels, it was intended that staff would respond to requests for guidance on sources and even arrange for remote access to the sources themselves through the use of photography and telephones adapted for the transmission of images.

Although the IIB was unique both in its scope and in the imaginativeness of its goals, it formed part of a wider movement in bibliography and information management characteristic of the late nineteenth and early twentieth centuries. A notable German initiative in the same vein as Otlet's was Die Brücke (The Bridge).[45] Founded in Munich in 1911, Die Brücke was dedicated to the organization of knowledge in ways that would enhance efficiency and, crucially, foster communication. Its chief inspiration was Wilhelm Ostwald, professor of chemistry at the University of Leipzig from 1887 until his retirement in 1906.[46] With Ostwald and the Swiss writer and publisher Karl Wilhelm Bührer as its joint chairmen, the Swiss journalist Adolf Saager as a leading collaborator and promoter, and Otlet a predictable choice as its honorary president, Die Brücke aspired to global reach as a mechanism for the diffusion of the fruits of intellectual endeavor of every kind. The essence of the project was organization. Books and journals were to be printed on sheets of standardized format and using approved typefaces. The prescribed paper sizes and uniform bindings in turn made for easy

storage and labelling on shelves adapted for the purpose (see figure 4).[47]
Once these various procedures were adopted, it would be possible to
advance to a union of all knowledge in what Ostwald conceived as a
"World Brain" accessible to the whole of humanity.[48]

4. Karl Wilhelm Bührer's ideal of a scholar's workspace, 1912. The space, no bigger
than a normal sitting room, offered easy access to sources of information, printed
in standardized formats and typefaces to facilitate information retrieval. The plan
formed part of the wider project of Die Brücke, the international institute for
the organization of intellectual work that Bührer and Adolf Saager founded in
1911, with the financial support of Wilhelm Ostwald. From *Bührer, Raumnot und
Weltformat* (Munich, 1912), 24.

Despite shows of interest by such prominent figures as Henri Poincaré,
Ernest Rutherford, the architect Peter Behrens, the Belgian industrialist
Ernest Solvay, and Albert I of Monaco, Die Brücke had a short life. It
was a costly venture, with ambitions that surpassed its means. Ostwald's
gift of much of his 1909 Nobel Prize for Chemistry, together with
hopes that Solvay's interest might translate into sustained financial
support, gave fleeting encouragement. But Die Brücke's precarious

existence ended in 1913 in bankruptcy and the demise of its fortnightly journal, *Die Brücke-Zeitung*. It left a modest legacy of elegantly printed pamphlets describing the project, though little else of lasting material value. Its significance, however, was of a different order. Like Otlet and La Fontaine's initiatives in Brussels, Die Brücke exemplified more than just the tide of internationalist sentiment in prewar intellectual life. The priority that Ostwald gave to easing access to knowledge was a means to the higher end of setting humanity on the path from the rhetoric of pacifism to the reality of a world at peace.

At the same time as the IIB was working toward a universal system of classification, most other bibliographical initiatives were targeting the needs of a particular discipline or professional community. In this respect, they were of a quite different stamp, with none of the overarching ideological overtones that motivated Otlet and La Fontaine. The engineer Hermann Beck's Institut für Techno-Bibliographie, founded in Berlin in 1908, was just such a venture, narrower in its scope than Die Brücke though still aiming to cover all forms of technical literature (perhaps too ambitiously, since the project soon died).[49] In the sciences, a more effective pacemaker in this focused approach to bibliography was the Royal Society in London. Faced with the inexorably accelerating pace of scientific publication, the society responded with exemplary efficiency to a proposal by the American physicist and secretary of the Smithsonian Institution Joseph Henry at a meeting of the British Association for the Advancement of Science in Glasgow in 1855. Henry's call was for a catalogue of scientific papers published across the world, and the result, more than a decade later, was a set of substantial volumes that are to be found in virtually all academic libraries to this day. A first series of the *Catalogue of Scientific Papers*, covering papers in all the sciences (though excluding technology) in the publications of scientific societies and other journals between 1800 and 1863, appeared in six volumes between 1867 and 1872. A further thirteen volumes, completing the coverage to 1900, were published intermittently until 1925.[50]

Although the *Catalogue* delivered on its promise, the growth and diversity of the periodical literature in science imposed a constant strain.

By 1894, the limitations of the original structure, especially the listing of items by authors' names and without distinction of disciplines, had led to a quite different plan, for a card catalogue of the world's scientific publications, classified by subject. It was this plan, conceived within the Royal Society and launched in a series of international conferences in London between 1896 and 1900, that took shape, in book form, as the *International Catalogue of Scientific Literature* (ICSL).[51] Under a distinguished international council and with almost thirty regional bureaus across the world, the ICSL was launched as a series of annual publications, one for each of seventeen areas of science, that covered papers published from 1901 to 1914, the last volumes (for 1914) appearing in 1921.

In turning from its initial unified coverage of the sciences to the discipline-focused bibliographies of the ICSL, the Royal Society was acknowledging the continuing trend to specialization in science. The ICSL was first and foremost a working tool that would be extensive in its seventeen subject areas and international in its coverage. But it too was soon seen as having its weaknesses, not least in the delay of up to a year before the references it recorded were available. By 1914, it was clear that the future of finding aids for scientists lay with publications that provided more rapid access to the literature than was possible through an annual listing. Substantial sections or supplements of abstracts in national disciplinary journals, such as the London-based *Journal of the Chemical Society*, the *Bulletin de la Société chimique de France*, or the *Zeitschrift für angewandte Chemie* in chemistry, represented a partial solution, as they had done since the mid-nineteenth century. But, as the challenge of achieving an acceptable level of coverage became more formidable, even those ambitious ventures fell short of the comprehensive scope that research at the cutting edge of any particular discipline required. Even the conspicuously well-managed weekly *Chemisches Zentralblatt*, the distant successor to the first of all abstracting journals, the *Pharmaceutisches Centralblatt* (launched in Leipzig in 1830), was thought to cover non-German material inadequately and to have some residual bias toward its original brief of covering mainly pharmaceutical publications.

Initially, *Chemical Abstracts,* the bold new departure that the American Chemical Society launched in 1907, had similar failings; it bore traces of its origins as the successor to Arthur Amos Noyes's US-focused *Review of American Chemical Research*, which had only noted articles published by Americans, though including those of Americans working abroad.[52] But, thanks to resolute management and unprecedented financial investment, by the eve of the First World War *Chemical Abstracts* could lay a plausible claim to completeness in its coverage not only of the world's chemical literature but also of patents. It was certainly not alone, nor was chemistry unique; *Science Abstracts*, launched at about the same time in London, fulfilled a comparable function for physics.[53] But *Chemical Abstracts* soon established itself as preeminent in the genre. By the time the paper edition ceased in 2010, it had published some thirty million abstracts, compared with about two million published in 140 years by the *Chemisches Zentralblatt*, which remained *Chemical Abstracts'* main rival until ceasing publication in 1969.[54]

Centers of intelligence

The more idealistic and predominantly nongovernmental forms of pre–First World War internationalism were expressed with such a diversity of emphases that characterization is difficult. A strong recurring streak of pacifism is beyond question, even though by no means all internationalists, Otlet included, were unqualified pacifists. More elusive, yet still significant, are signs of incipient feminist sympathies, the first stirrings of socialism, and challenges to colonialism and slavery, seen as epitomizing one people's or one race's exploitation by another. There are also pointers to a kinship with certain aspects of late-nineteenth-century positivism, distantly inspired by the ideas of Auguste Comte (who had died in 1857) but given new life in late-nineteenth-century independence movements in Latin America and the secular policies of the determinedly secular Third Republic in France. Among the idealists, the exposure of Otlet in particular to positivism is beyond question:

his belief in the sciences as a force for social and intellectual progress had much in common with that of his positivistically inclined Belgian contemporary and historian of science, George Sarton.[55]

Although no single profile fits everyone, one thing that united prewar internationalists, including many of those speaking in the name of governments, was a belief in the value of association and communication across national, linguistic, and cultural boundaries. It was a belief that offered a bedrock on which good human relations could be constructed and the essential oneness of humanity expressed. Among leading scientists whose internationalism bore this stamp, Wilhelm Ostwald stood out. While the aspirations of Die Brücke to encompass all branches of knowledge had much in common with those of Otlet and La Fontaine, Ostwald also engaged with the mechanisms of communication and access to knowledge within his own disciplinary community, in chemistry. Here, as in his broader enterprises, his leadership role was recognized. When the International Association of Chemical Societies (IACS) was established in April 1911 at a meeting in Paris of representatives of the national chemical societies of France, Britain, and Germany, Ostwald was a natural choice as president.[56] Within the IACS, his presidency of the biggest European chemical society, the Deutsche Chemische Gesellschaft, gave him special authority. But he worked effectively with the other main protagonists, Albin Haller (for the Société chimique de France) and William Ramsay (for the Chemical Society in London), in what all three men saw as an important common initiative.

The IACS responded to a widely felt need for integration among national communities in the discipline, and it quickly established itself as an internationally recognized focus for the day-to-day management of conventions on nomenclature, chemical symbols, physical and chemical constants, and other matters of concern to practising chemists, including the ever-present task of information management and retrieval. By the time it held its second meeting, in Berlin in April 1912, societies representing the United States, Holland, Russia, Denmark, Austria, Spain, Italy, and Switzerland had joined, and other national communities were seeking affiliation.[57] At the third meeting, in Brussels in September

1913, delegates from fourteen national societies (representing a total of almost twenty thousand members) were present, and the IACS had the prospect (eventually unrealized) of an initial donation of 250,000 francs, with long-term support to follow, from Ernest Solvay.[58]

For a man of Ostwald's determinedly cosmopolitan spirit, even this degree of federation was not enough. In parallel with his presidency of the IACS, he fashioned a scheme for an international institute of chemistry that would provide a global center for the diffusion of chemical knowledge but also have laboratories and facilities for a core resident staff as well as visitors from all over the world, regardless of nationality.[59] The neutral status of Belgium made Brussels the natural location for the institute (ideally with an associated center in the United States), and it was there that Ostwald planned to deposit his library of four thousand volumes and twelve thousand offprints as the nucleus of a "universal" library of chemical science.[60] Despite Ostwald's personal generosity, including the promise of a donation from his 1909 Nobel Prize, further funding would be needed. For this, hopes rested again on Solvay, who had already helped the IIB and who now promised a substantial sum for the institute in tandem with his support for the IACS. In the event, business pressures (probably reinforced by second thoughts about the soundness of the proposal) soon led Solvay to withdraw his offer.

Solvay's loss of interest was a disappointment for everyone involved in the IACS, which began to falter, despite many declarations of sympathy, before the war precipitated the suspension of its activities. Ostwald, of course, was particularly affected, as he was by the failure of the plan for his institute and the tribulations of Die Brücke, which together exemplified the difficulties that beset international projects, however great the commitment of their promoters. Yet his resolve never weakened. This was nowhere more evident than in his concern for language, which he saw as essential to any internationalist program. His own background as the child of German parents in Latvia, his education and early research in Riga and German-speaking Dorpat in Estonia, and his professorial career at the Universities of Riga and Leipzig left him with a competence in French, English, Russian, and Latin, as well as German, that allowed him to surmount most linguistic barriers with ease. But his recognition that such barriers worked against his ideal of an unimpeded circulation of knowledge left him a natural champion of some form of constructed language that could be used as an auxiliary of "ethnic" languages for scientific purposes.[61]

In his interest in linguistic innovation, Ostwald stood in a tradition going back to the pioneering efforts of Bishop John Wilkins and G. W. Leibniz to establish a "philosophical language" in keeping with the new science of the seventeenth century. But what had then been cast as a rather distant dream had now become a matter of urgency. By the later nineteenth century, German, English, and French, in roughly descending order, were all widely used for scientific purposes, with Russian as a receding outlier.[62] Although, in principle, the adoption of any one of them as an approved standard in conferences and publications would have been feasible, the certainty of the friction that such a choice would have provoked made the alternative of an auxiliary language attractive.[63]

Among a number of such languages that had their champions, the most successful was Esperanto (Esperanto for "hopeful").[64] It was the work of Ludwik Lazarus Zamenhof, a Polish eye doctor and campaigner for international peace and understanding very much in the mold of Otlet and La Fontaine.[65] The language's main foundations lay in the Romance languages; its syntax was Romance, and two-thirds of the vocabulary was of Romance origin. But its simplified grammar made it accessible to those beyond the Latin countries (albeit far less so to those with a non-European linguistic background). Between 1887, when Zamenhof published an outline of his scheme, and 1905, when his *Fundamento de Esperanto* laid down the definitive principles of grammar and vocabulary, the language caught on, especially in Russia and central Europe. By the time of the eighth annual Universal Esperanto Congress in Cracow in 1912, some fifteen hundred Esperanto societies existed across the world (though mainly in Europe) and the movement laid claim, perhaps optimistically, to hundreds of thousands of users.[66]

A predecessor and initially a serious rival to Esperanto was Volapük ("worldspeak" in Volapük), promoted from 1879 by a German Catholic priest, Johann Martin Schleyer.[67] By 1889, the two hundred people who attended the third international Volapük convention in Paris conducted all their business in the language. But destructive schisms between users, an unfamiliar vocabulary, and complex grammar soon weakened the cause, provoking defections to Esperanto and setting Volapük on a downward path from which it never recovered. A stronger competitor, especially as a vehicle for the sciences, was Ido. With roots in a reformed Esperanto, Ido was created in 1907.[68] It was Ostwald's choice, to the point that he advocated its use in the plan for his international institute of chemistry and again dug into his 1909 Nobel Prize in support of the cause. Emerging as it did at the height of the prewar internationalist movement in science and scholarship, Ido did not want for prominent advocates. Notable among these was the French mathematician and philosopher Louis Couturat, who had championed the quest for an international language since witnessing the difficulties of communication during the congresses that accompanied the Exposition Universelle of 1900 in Paris. While Couturat accepted that the new language would be "auxiliary" in the sense that it would complement and not replace vernacular speech, he saw it as a solution to the linguistic diversity that presented, in Couturat's words, "an immovable barrier to a complete understanding and to an intimate insight into one another's minds."[69] It was axiomatic for Couturat, as for Ostwald, that from understanding and insight there would flow "peace and concord."[70]

Advocates, though, were not enough. In a manner characteristic of the auxiliary languages, Ido too was to suffer from factions that undermined the promoters' best efforts, and the war, coming hard on the heels of Couturat's death in a road accident in 1914, dealt it a serious blow. Nevertheless, like Esperanto and Volapük, Ido had its brief heyday. Otlet and La Fontaine were natural sympathizers, and they embraced the advocates of the various languages in their plans. The ideal of linguistic universalism embodied the same transnational aspirations that drove their own ventures, and although they never learned any of the languages, they attended Esperanto congresses and publicized both Esperanto and Ido in their lectures and writings.

A metropolis of communication

In other areas too, Otlet and La Fontaine did not have to look hard for kindred spirits. They found one of the most committed and eloquent of them in Patrick Geddes, professor of botany at University College Dundee (1888–1919) and then of sociology at the University of Bombay (1919–1924). Geddes's interest in urban improvement, with its roots in his evolutionary interpretation of human history and his concrete plans for a renewed postindustrial Edinburgh, made him a natural sympathizer in any proposal that might lead to a happier humanity.[71] His cultural cosmopolitanism too played its part, cementing the bonds of friendship and reciprocal admiration between him and Otlet. The correspondence that the two men exchanged from their first meeting in Paris at the Exposition Universelle in 1900 until 1930, shortly before Geddes's death, bore witness to a rare meeting of minds.

While Geddes's writing had a visionary quality that complemented his practical concerns, it never conveyed the extremes of idealism that fired the "Super-Metropolis" (Geddes's term) of his contemporary Hendrik Christian Andersen's plan for a community that would live the realities of a universally interconnected world. Andersen was a Norwegian-born American artist, aesthete, and city planner now best remembered for his close, largely epistolary friendship with Henry James between 1899 and James's death in 1916 and for the museum of sculpture and other art (much of it his own) in the Villa Helene, which he built close to his longtime residence in the piazza del Popolo in Rome.[72] In the plan for his ideal community, Andersen set out to create an urban setting for a utopian vision rooted in the powerful streak of pacifism that sustained his hopes for a better world.[73]

The guiding principle of his city was that it should be a "World Centre of Communication" dedicated to facilitating exchanges, and hence solidarity, between peoples and nations. It was through this transcending of cultural and political boundaries that the city would fulfill its still higher mission of helping humanity on its way to the state of moral, physical, and intellectual perfection that Andersen regarded as its proper destiny.[74] Andersen's personal cosmopolitanism was essential to the whole project. This was bred of an upbringing in Rhode Island, to which his Norwegian parents had immigrated when he was an infant, and extensive European travels culminating in his residence in Rome, where he lived from the mid-1890s until his death in 1940. It was as a man of many friendships that he was able to count on the support he received from both friends and wealthy patrons, among them his sister-in-law and collaborator Olivia Cushing. That support, allied to his talents as a sculptor and publicist, gave his project the status and plausibility it required. It allowed him to devote nine years to formulating his plan, with the support of a team of forty architects, including his closest associate, Ernest Hébrard, an accomplished French architect and urban planner whose technical skills complemented Andersen's imaginative flights.

The fruit of the collaboration appeared in 1913 in sumptuous privately printed English and French expositions of the project, preceded by a lengthy essay on the history of city planning from the earliest times by the professor of classical archaeology at the University of Bordeaux, Gabriel Leroux.[75] Modern techniques for all forms of communication, integrated with a provision for comfort and the pleasures of life on a human scale, were at the heart of the plan. To the essential end of international harmony (though not the elimination of national identities), embassies and national cultural centers would be clustered along a central Avenue of Nations (see figure 5). Around the center there were to be residential quarters, six in all, each housing a hundred thousand people. In this way, residents would live in communities sufficiently small to allow for personal interactions but large enough in aggregate to justify the city's "world" status. The same dual purpose characterized the facilities for recreation. Gymnasia and swimming pools were for public use, and an artificial lake, the "Natatorium," would allow all citizens to engage in yachting and other aquatic sports.

The eight-hundred-meter-long stadium and the adjacent "Centre for Physical Culture" similarly offered an opportunity for citizens'

5. Statues symbolizing peace at the entrance to Hendrik Christian Andersen's ideal World City, 1913. The statues represented manhood and womanhood "in their highest physical development." Framing the city's central Avenue of Nations, with their torches serving as lighthouses, they were to rise to a height of eighty meters. Image from Rome, National Gallery of Modern and Contemporary Art, Museo H. C. Andersen; also published as a frontispiece to part two of Andersen, Creation of a World Centre of Communication (Rome, 1913). Image from Rome, National Gallery of Modern and Contemporary Art. Museo H. C. Andersen. By permission of Ministero dei Beni delle Attività Culturali e del Turismo, Rome.

personal fulfillment. But the stadium was also planned to promote the internationalist cause by serving every four years as a permanent home for the Olympic Games.[76] Andersen was well aware that the initial conception of the Olympic movement as an instrument of peace, which had helped to fire Pierre de Coubertin's plan for the resurrection of the ancient games, the Games of the I Olympiad in Athens in 1896, was not always matched by the reality. French reticence at the admission of German athletes soon laid bare the intrusion of international politics, with roots in lingering enmities going back to the Franco-Prussian War. And all too easily, athletic success or failure fed unhelpful perceptions of national superiority or decline. The Fifth Olympiad, in Stockholm in 1912, provoked just such concerns, with national interests, including Sweden's aspirations to the status of a major power, much in evidence; Svante Lindqvist's comment that the Stockholm Games had something of the character of a "dress-rehearsal for the First World War" rings disquietingly true.[77] If such nationalistically motivated competitiveness had any effect on Andersen, it was only to heighten his idealism. In this, as in everything he did, he aspired resolutely, even blindly, to all that was best in human nature, achieved by individuals rather than by nations. Hence it was Coubertin's ideals, rather than the departures from them, that mattered to him.

As an exercise in innovative urban planning marshalled to promote living on a human scale, Andersen's World Centre had much in common with what Robert Kargon and Arthur Molella have described as the "invented Edens" of the twentieth century and more specifically with the notion of the Garden City that Ebenezer Howard had elaborated a decade earlier in his *Garden Cities of Tomorrow* (1902).[78] But the priority that Andersen gave to peace and the defusing of conflicts, with shared understanding as essential to the process, set his city apart from other "Edens." There was to be an international bank and court of justice, an international reference library, and a "temple" to provide meeting places for the pursuit of religion (always, significantly, with the emphasis on the common ground between religions, rather than on the differences between them). And most importantly, at the heart of the city there were to be four monumental congress buildings devoted to debate and the exchange of information about science and its applications, including one for the "sociological sciences."

Although Andersen was no scientist, his perception of science as the field of knowledge best attuned to his vision was explicit. In his view, science had an intellectual and even moral superiority that allowed it to transcend boundaries and strife. At the root of this perception was a belief in the purity of science. In this, the analogy with the benefits of physical exercise was clear and deliberate. Just as strenuous effort and training were the keys to physical well-being, so the practice of science— essentially the exercise of humanity's universal capacity for reason and observation—offered the surest path to the health of the mind and the moral qualities that went with it. The goals of bodily health and mental perfection were, in fact, linked inextricably, and could be achieved only through a process of elimination, whether of illness or of error. Speaking of science in his *Creation of a World Centre*, Andersen played strongly on the metaphor of purification:

> It is in the power of science to purify the world, to exterminate destructive germs from every nerve and fibre, to give strength and precision to all mental and physical efforts; and it is our privilege to be born in an age when all mankind, drawn more closely together, seems to be desirous of advancing on the highest and most beneficial lines for the general good.[79]

As well as standing as a symbol of purity, science was also to have a more concrete function, as the fount of the technologies on which the exchange of information in the modern world depended. The Tower of Progress, dominating a vast square in the city's "scientific centre," expressed the excitement and promise of science-based modernity (see figure 6).[80] At 320 meters, it was conceived as an engineering feat in itself. It was to be twenty meters taller than the Eiffel Tower, and its more than forty stories were to house state-of-the art installations for communication of every kind: an international news bureau, and

6. Tower of Progress, a 320-meter-high focal point of Hendrik Christian Andersen's World City. The tower was to stand at the heart of the city's "scientific centre," surrounded by monumental buildings devoted to the hosting of international congresses and other cultural activities The tower itself housed facilities for many aspects of communication, including wireless transmission, accommodation for international societies, and an international news bureau. Image from Rome, National Gallery of Modern and Contemporary Art. Museo H. C. Andersen. By permission of Ministero dei Beni delle Attività Culturali e del Turismo, Rome.

facilities for wireless telegraphy, the printing of newspapers and journals, the reception by wireless of lectures in nearby congress buildings, and a system of pneumatic tubes that would allow books to be transported from and between libraries, all connected to the central hub of a comprehensive "international reference library." Beneath the tower, a basement would give access to three stations of the city's underground railway system in a vast underground concourse.

Among the contemporaries who saw the World Centre as an unrealizable fantasy was Andersen's dear friend Henry James, who wrote candidly urging him to return to the path of "sound & sane Reality" and to beware "the dark danger of Megalomania."[81] James's pain at the very thought of what he saw as Andersen's monumental aberration ran deep: "The unutterable Waste of it all makes me retire into my room & lock the door to howl!" Andersen, though, was undeterred. With no apparent regard for cost, he sent copies of his book, many of them with a flyleaf specially printed for the recipient and often with handwritten dedications, to major libraries throughout the world and to chosen individuals he identified as actual or potential sympathizers.[82] In this way, abetted by a mixture of ephemeral publicity literature and personal approaches to people of influence, he accumulated a formidable dossier of endorsements. Not everyone responded as extravagantly as the French astronomer Camille Flammarion, who saw the World Centre as a "magnificent idea" that would advance his own dream of a humanity liberated from its "animal chrysalis" and the Earth's "ancestral barbarism" to live as free spirits, citizens not of nations but of the whole Earth, even of the heavens.[83] But praise and support came from across the spectrum of progressive thought. Ostwald, John Lubbock, and Melvil Dewey were among the hundreds of correspondents who took the trouble to express their approval, with the more zealous of them enrolling as members of the World Conscience Society that Andersen established to promote his ideas.[84]

From the start, Otlet and La Fontaine were among the World Centre's most enthusiastic supporters. Even before Andersen's book appeared, they had examined the drawings in Hébrard's Paris office and welcomed the project as the "architectural realisation" of their own "functional activity."[85] It is easy to see why they found so much to admire in Andersen's visionary thinking. To critics, especially those who saw gathering international tensions as a harbinger of inevitable war, the idea that the World Centre might avert the looming conflict appeared a pipedream. But Otlet and La Fontaine saw the plan very differently. For them, it was an endorsement of their aspirations and evidence that the trend of world affairs, exemplified in the Peace Conferences of 1899 and 1907 in The Hague, was running in their favor.

Encouraged by what they saw as an accumulation of favorable circumstances, Otlet and La Fontaine gave new priority to providing an institutional structure within which their various initiatives could be coordinated. The main vehicle for this was the Union des associations internationales (UAI), which they were instrumental in founding in

Brussels in 1910. The function of the UAI and its executive wing, the Office central des associations internationales (OCAI), was to unite the efforts of nongovernmental bodies with internationalist goals across the world. At the inaugural Congrès mondial des associations internationales in Brussels, during which the UAI was formally constituted, with Otlet and La Fontaine as joint general secretaries, 132 of these were represented.[86] A second, highly successful congress followed in 1913, also in Brussels, this time with 169 bodies represented from twenty-two nations. By now, an annual handbook (*Annuaire de la vie internationale*) and a substantial monthly journal, *La Vie internationale*, ensured continuity and publicized the UAI's activities, in particular its role as a coordinating body for international congresses; more than 140 of these were listed in *La Vie internationale* for 1912, almost a third of them on areas related to science, technology, and medicine.[87]

By the eve of the war, the UAI was riding high, with 230 societies and other organizations affiliated to it. Crucially too, the annual funding that Andrew Carnegie provided through to 1914 gave it not only the prospect of enduring prosperity but also the cachet of an association with one of the world's greatest champions of peace. The effect of Carnegie's patronage could scarcely have been greater. The Carnegie Endowment for World Peace, launched with Carnegie's donation of $10 million, was established in the same year, 1910, as the UAI, and it embodied the same confidence that understanding would be best achieved through exchanges between the leaders of intellectual life, irrespective of nation.[88] Between Carnegie and the leaders of the UAI there was a true meeting of minds.

With so many elements converging to promote exchanges and collaboration across frontiers by 1914, Otlet and La Fontaine had reason to believe that the causes in which they were involved were poised for even greater success. The internationalist spirit seemed to be approaching its zenith. Intergovernmental bodies were playing a significant part, albeit with varying degrees of effectiveness and commitment, and invariably with a watchful eye on national interests that might be jeopardized by an excess of international sentiment. But the movement's real strength lay in independent communities and organizations. While the UAI encapsulated the spirit of this strand of internationalism, it did not represent the sum of it. Scientists and disciplinary bodies too set a brisk pace with their congresses and working parties on nomenclature, units, and standards, as did the International Association of Academies, playing to its particular strength as an association of formally constituted academic elites.

In such a diverse and crowded movement, the UAI had to work hard for recognition of the leadership or at least the coordinating role to which it aspired. Finances too were a concern: dependence on personal donations such as those from La Fontaine and Carnegie, rather than on sustained public support or an accumulated endowment, made for insecurity. But the UAI and those who shared its ideals never flinched in their belief in the free circulation of knowledge as the best antidote to misunderstanding and conflict. It was a belief unequivocally endorsed by internationalists on both sides of the divide that was soon to separate them into hostile camps and challenge the very idea of the unified cosmopolitan world of learned culture to which, on the very brink of war, they still subscribed with such optimistic enthusiasm.

CHAPTER TWO

War as Watershed

On the eve of the First World War, no one could be unmindful of tensions that bore the potential to transform local conflicts into more general warfare. The fragility of Franco-German relations since the war of 1870, bitter colonial rivalries in North Africa, and finally the Balkan wars of 1912–1913 gave mounting cause for alarm. Yet the experience of a century without war on anything approaching a European scale and the institution of procedures for settling disputes peaceably calmed many anxieties. Special hopes were invested in the Peace Conferences in The Hague. These lengthy affairs had resulted in a series of final acts and declarations of intent aimed at banning certain weapons (including gases), controlling the use of others, and reducing or at least limiting the stocks of arms held by the major powers. It is true that, in the event, the conferences achieved less than had been hoped; translating good intentions with regard to conflict resolution, disarmament, and the restricting of military budgets into binding conventions proved a daunting task, and governments hesitated to commit themselves to the agreements that the conferences called on them to ratify. Nevertheless, the twenty-six delegations that had been present in 1899 and the forty-three attending in 1907 had come together in a quest for lasting harmony, or at least stability, among nations, however differently they might conceive the best route to that end.

In their aspirations, the participants in the conferences shared much common ground with champions of the internationalist cause in science, including some of the most prominent scientists of the day. Until the very eve of the war, these champions clung to the belief that the global networks they were striving to promote—and the congresses, personal correspondence, and publications on which the networks depended—would act as a bulwark against major conflict. But as what many in 1914 thought would be a short-lived confrontation lengthened into protracted stalemate, their expectations collapsed. For science, the war was a watershed, one whose course and immediate aftermath, extending into the early 1920s, did irreversible damage to any notion of a seamless web of learned culture capable of rising above politics and national interest.

Science at war

The assassination of Archduke Franz Ferdinand of Austria in Sarajevo on June 28, 1914, and the succession of declarations of war in late July and early August dealt a blow that shattered the tranquil world of learned culture, as it shattered individual lives and transformed economies. Nowhere were the consequences felt more poignantly than in Belgium, whose neutrality the German army violated with impunity on August 4. Faced with the rapid occupation of all but a small part of their country, Belgians who were in a position to flee had to make a quick decision. George Sarton, commonly regarded as the founder of the history of science as a discipline, left Ghent for Holland, then England, and finally the United States, where he made an academic career for himself, mainly at Harvard, with the support of the Carnegie Institution of Washington. Otlet too chose the path of exile. After a short stay in Holland, he made his home in Lausanne, leaving the Institut international de bibliographie to continue its work as best it could under its secretary, Louis Masure. From Switzerland he embarked on a campaign of a different kind, for a world charter that would guarantee peace and human rights under the aegis of a confederation of states of the kind that eventually took shape as the League of Nations.[1] La Fontaine ventured farther afield, though in the same missionary spirit. He went first to London in September 1914, then in the following April to the United States, where he pursued his ideals, lecturing on international relations and campaigning for a union of humanity in accordance with what he called his "Magnissima

Charta." In this draft declaration, he set out the rights and duties of states in what he advanced as his "Great solution" in the cause of peace.[2]

For practising scientists, escape was less easy, and in all the warring nations scientific communities found themselves swept up in the patriotism that bound them to their country's cause. As the old internationalism withered, the frequently quoted assertion of the German chemist Fritz Haber—that in time of peace the scientist "belongs to the World" whereas in war he "belongs to his country"—expressed a common view.[3] But the harnessing of science in the cause of destruction on an unprecedented scale lent new and more brutal significance to the notion of the mobilization of scientists as classically exemplified in the French revolutionary wars of the 1790s. Improved explosives, machine guns, and aerial bombardments all contributed to the toll of science-based annihilation inflicted on combatants and noncombatants alike.[4] Among the weapons, poison gases assumed a special status from the moment they were first used near Ypres in April 1915. It was not so much that the gases caused death, although they did, albeit in a minority of cases and often only long after the victims had left the battlefield.[5] Their particular effect was psychological. And it was their capacity to spread fear and to do so as a virtually invisible agent that made Haber an object of particular detestation in the Allied countries.

The cherished conception of science as a form of knowledge that knew no national boundaries was equally vulnerable to the abuse that the scientific communities on the two sides of the conflict exchanged throughout the hostilities. The exchanges between intellectuals began promptly; and, once started, patriotically inspired insults flowed freely in both directions.[6] Just a few weeks into the war, vilification reached fever pitch when a group of ninety-three German intellectuals published a patriotic manifesto under the title "A Call to the Civilized World."[7] The Manifesto of the Ninety-Three, as it became known, published in the German press on October 4, 1914, and quickly translated and circulated abroad, was a vibrant expression of national pride and resentment against the Allied powers, all cast in an unyielding tone that was to color perceptions of Germany long after the Armistice of 1918.

The essential contention of the ninety-three was that the Kaiser had never wanted the war. Germany, though, had had no choice but to defend itself against the "lies and calumnies" with which the Allies had endeavored to "stain German honour." The manifesto went on to enumerate the calumnies and answer them one by one. To the charge that Germany had violated Belgian neutrality the answer was unequivocal: Germany had simply acted to forestall British and French plans to occupy the country. Then, in response to one of the Allies' gravest accusations, could it really be claimed that the German army's action in destroying much of Louvain had been gratuitous? The firing of parts of the city had been a necessary response to the attacks by Belgian citizens on German soldiers, and if the church of St. Pierre and the university library had been burned, that was no more than justified reprisal. The Allies might see things differently and say all they liked about German "militarism" and supposed atrocities against the civilian population. But how else was Germany to defend its culture? This was, as the manifesto had it, a war of cultures in which the Allies had set out to target the very "soul" of the German people.

The distinction of those who signed the manifesto left no doubt as to Germany's standing in science and scholarship. Of the scientists (who made up about a fifth of the signatories), Einstein, a pacifist and internationalist, was the only significant absentee from a list that included (among past and future Nobel Prize winners) Fritz Haber, Wilhelm Ostwald, Adolf von Baeyer, Emil Fischer, Paul Ehrlich, Max Planck, Phillip Lenard, Walther Nernst, Wilhelm Wien, Richard Willstätter, Wilhelm Röntgen, and Emil Adolf von Behring. A document that bore such names could not be allowed to pass unanswered, and commentators in the Allied countries soon showed themselves quite as adept in defamatory analysis as their German peers. Despite the years of respectful admiration that had gone before, the aim on the Allied side was now to demonstrate that the reputation of German science was exaggerated.

A favored Allied assertion was that German scientists privileged quantity over creativity and subtlety. The British chemist William Ramsay laced his disparagement with a racial edge in the journal *Nature*. According to

Ramsay, the greatest advances in science had not been made by members of the German race, the "Teutons," who were not to be confused with the "Hebrews resident among them": indigenous German science was derivative and had yielded only a "deluge of mediocrity."[8] Pursuing the idea of a distinctive German style of science, the French mathematician and philosopher Pierre Duhem drew on Pascal in contrasting between the "esprit de finesse," which sprang from a refined experience of the world, with the "esprit géométrique," founded on logic and abstraction.[9] The former was the French way to understanding; the latter, German way, accounted for the obscurity and prolixity of German science, made worse by the heaviness of the German language.

Counter-affirmations of the high quality of German science were not wanting. From Germany's nationalistic political right, the physicist Wilhelm Wien expressed a characteristic mixture of conservatism and nationalistic fervor in circulating a declaration (his "Aufforderung") in which he called on his fellow physicists to shun publications in English by refusing to cite them; words of foreign origin that had crept into German usage, such as "Skineffekt" or "Äquipartitionsgesetz" should also be avoided.[10] In an atmosphere of such mutual hostility, the rare voices of moderation went virtually unheard. When in mid-October 1914 the German physiologist and pacifist Georg Friedrich Nicolai responded to the Manifesto of the Ninety-Three by circulating a petition urging "Europeans of education and good will" to exert their influence in the cause of peace, he found only three colleagues, all fellow professors in the University of Berlin, willing to sign: Einstein, the professor of astronomy and former head of the Berlin Observatory Wilhelm Förster, and the philosopher Otto Buek. Wartime German censorship did the rest, and the petition was lost from view until it reemerged in 1917 in the preface to Nicolai's *Die Biologie des Krieges*.[11]

The alacrity with which the competing stereotypes were articulated is a mark of the prejudices from which they sprang.[12] The prejudices ran deep, especially in France and Germany. The French had long seen their science as the product of a refined Latin mentality far removed from the efficient but plodding culture of the north, whereas the Germans had prided themselves on the nineteenth-century refashioning of their universities as centers for specialized research and publication. Adding force to the contrast were the divergent traditions of Catholicism in the south and the Protestantism of Prussia. Confronted with such cleavages, notions of a unified world of learning that had been de rigueur, at least on the surface, only weeks earlier suddenly appeared as vain rhetoric. Plainly, the normal procedures of international science could not continue, and with memories of the ninety-three continuing to poison relations, formal contacts between scientists of the warring nations ceased.

In learned societies, the conflict between patriotic duty and loyalty to the wider scientific community provoked debates on the wisdom and morality of retaliatory action against members in enemy countries, with different results. In Britain, the Chemical Society took a belligerent line, deleting the names of all German honorary members from its lists in 1916, following protracted discussions. The Royal Society and the Paris Academy of Science, on the other hand, decided to take no action, although toward the end of the war a proposal that the Royal Society should remove "enemy aliens" from the category of foreign members did receive some support.[13] In Berlin, the decision on exclusion was a close-run thing. But Planck and Fischer managed to persuade the Prussian Academy of Sciences to stop short of severing all ties with French academies.[14]

From the moment hostilities began, most international ventures in science were either abandoned or cut back to a perfunctory level. Such bodies as the International Association of Chemical Societies, barely three years old in 1914, and the International Association of Academies had no function in time of war, and they survived as purely administrative entities until their final dissolution in the course of the three conferences of the leading Allied academies between October 1918 and July 1919 (on which see below). The interruption of normal cultural relations affected neutral nations as well as the parties to the war, though without stopping all activity. In the United States, which did not enter the conflict until April 1917, a state of planning that was already well advanced when hostilities began allowed the Panama Pacific Exposition in San Francisco to open in February 1915 and to run for nine months. In that time, the exhibition attracted between sixteen and eighteen million visitors, for whom the exhibition passed off much as it would have done in peacetime.[15] Most striking was a strong representation from abroad, including major displays from twenty-nine countries, most of them in national pavilions built for the occasion. Despite the war, and in contrast with a low-key German and British presence, France was much in evidence. Its pavilion, a replica of the elegant eighteenth-century building in Paris that had housed the headquarters of the Legion of Honour since the order's creation under Napoleon, contained an exhibition area larger than that of any other nation. Although Rodin's bronze figure of *The Thinker* before the entrance caught most eyes, learned culture too had its place, notably in a "Bibliothèque de la science française," an ambitious book exhibition accompanied by two volumes of essays chronicling French achievements in the different areas of science and scholarship.[16]

In the Scandinavian countries, determinedly neutral but close to the conflict both geographically and through cultural and diplomatic ties, especially with Germany, the challenge to normality took a different form. The Nobel Prizes raised particular difficulties. The fact that the decisions about the three science prizes—for physics, chemistry, and physiology or medicine—and the prize for literature were made in Sweden, with the Norwegian parliament awarding the Peace Prize, meant that there was no formal reason why the customary procedures should not continue. In August 1914 members of the sectional committees for that year's prizes duly began their deliberations. But the realities of war soon prevailed. The danger that the award of a prize to a citizen of a belligerent country might be construed as a mark of sympathy for the nation in question, and hence as a retreat from true neutrality, weighed heavily. For this reason, the committee for the chemistry prize (despite making its choice, of the American Theodore W. Richards) broached the possibility of a one-year postponement of the award, and in November 1914 the Swedish government determined that while the decisions for 1914 could be made and announced, the presentation of the prizes should be put off until 1915.[17]

The postponement was controversial. What message did it convey about the Nobel institution's will to rise above the turmoil of world affairs? Within Sweden, nationalistically inclined conservatives, in particular, saw it as a pusillanimous capitulation that betrayed Nobel's intentions and undermined the country's special status as a bastion of uncorrupted neutrality. For their mainly liberal opponents, on the other hand, postponement stood as an affirmation of the impartiality essential to proper decision-making; their view was that war made such impartiality impossible, leaving suspension as the only option. With all hope that the conflict might be brief now abandoned, the presentation ceremony that had been suggested for 1915 did not take place, and the upshot, for the rest of the war, was compromise. The committees in the various sciences functioned rather fitfully, considering any nominations they could muster and either drawing up partial slates of winners or declaring that no award would be made for the year in question.

Such slates as were made public during the war years were not undistinguished. Far from it. They included Max von Laue (1914) and the Braggs, father and son, in physics (1915), and, following Richards, Richard Willstätter in chemistry (1915). But the irregularity of the awards bore witness to the profound disruption of a system that had encountered the impossibility of properly honoring Nobel's wishes. In a world haunted by the spectacle of unimaginable carnage and by an awareness of the scientific foundations on which so much of the carnage rested, the prizes' peaceful objectives were quickly lost from view. Under the pressure of circumstance, the decision not to hold the traditional Nobel ceremonies and to allow the committees to dispense with naming winners if they so wished became a recognition of the inevitable.

The phony peace: Storms over Stockholm

The Armistice of November 1918 ushered in a new climate for international initiatives. It did not do so, however, in ways calculated to promote harmony and reconciliation: four years of mutual recrimination and suspicion had taken their toll, leaving the war to cast its shadow on intellectual life for some years to come. The Nobel Prizes in particular provided an immediate battleground in which lingering enmities challenged the Nobel norms of peaceful emulation, with its principle that "no consideration whatever be given to the nationality of the candidates."[18] A central difficulty besetting the Nobel institution was a decision-making process that almost invited dissension, whether between competing national interests or of a more personal kind. Among several layers of complexity was the mechanism for proposing names, with its built-in weighting in favor of nominations from Scandinavia. While former prize winners and a small number of professors outside Scandinavia, whatever their nationality, had the right to make nominations for the prizes in physics and chemistry, so too did a far larger group composed of the home and foreign members of the Royal Swedish Academy of Sciences and professors in a number of designated Nordic universities.[19] A similarly strong Scandinavian presence colored the nominating procedures for the other scientific prize, in physiology or medicine. And the Scandinavian voice became even more pronounced at the next stage in the process, when the disciplinary committees appointed by the academy or (for physiology or medicine) the prestigious Karolinska Institute medical school in Stockholm chose the winners from the nominations before them.

Even in the limited circles of the committee members who made the choices, consensus had never been easily achieved. One unresolvable conflict in the prizes' early years had presented a major obstacle to Nobel's ideal of judgements delivered in accordance with criteria that had universal assent. The main antagonists in the conflict were the physical chemist Svante Arrhenius and the mathematician Gösta Mittag-Leffler. In personality, career profiles, and intellectual interests, the two men were very different. By contrast with the outgoing, well-connected Arrhenius, Mittag-Leffler was reserved and often ill, to the point of being an almost perpetual invalid; in so far as he had sympathies beyond the narrow world of Swedish science, these tended to be directed to Germany and secondarily to France.[20] More important in its consequences was a disparity between the two men's judgements of what constituted good science, with Arrhenius systematically preferring candidates whose achievements squared with his own work as an experimentalist, and Mittag-Leffler predominantly favoring theorists. It may well be that a lack of consensus on this fundamental point accounts for the failure of the chemistry committee to recognize the claims of the German physical chemist Walther Nernst until 1920, despite repeated nominations and powerful support from the international community. While Nernst's case may have suffered from the diversity of problems on which he worked, the theoretical cast of his research must have counted against him in Arrhenius's eyes and hence on the committees for chemistry and physics, either of which might have chosen Nernst as a worthy recipient of a prize.[21]

This and some other odd outcomes, such as the controversial award to Dalén in 1912,[22] had fostered a legacy of contentiousness that left the Nobel Prizes vulnerable to criticism. In the circumstances, as Elisabeth Crawford has argued, it is to the Nobel institution's credit that most prewar awards had broadly accorded with international evaluations of the work being recognized.[23] But a sequence of decisions taken immediately after the war took contentiousness to new levels. Allied unease had already been primed by suspicion that the manner in which the Nobel committees had asserted their aloofness from the conflict concealed the sympathy for the German cause that was known to be widespread in Sweden, despite the country's formally neutral position. With the war over, however, suspicion now turned to indignation as the Swedish committees stood firm against the vindictiveness of the Allied scientists' attitudes toward their German and Austrian peers. In taking their principled stance, the committees presented themselves as acting in accordance with the best traditions of the prizes and upholding the very internationalist ideals that the Allies were intent on betraying. The Allies, for their part, saw nothing but the antithesis of those ideals in a perverted high-mindedness that resulted in the award of five out of the seven Nobel Prizes in physics and chemistry between 1918 and 1921 to German recipients.

Most damaging of all for the Nobel ideal of calm decision-making was the award of the 1919 prize for chemistry to Haber for the synthesis of ammonia from its elements, hydrogen and atmospheric nitrogen, made industrially viable in the Haber-Bosch process.[24] The award had a history. Haber had been nominated in 1916 (when no decision was made), and he had returned to the list of candidates for 1918. In what was admittedly a thin year for nominations, he was nominated just once, by the Munich chemist Wilhelm Prandtl. The case, though, was strong, and it presented the chemistry committee with a dilemma. The committee's chairman, the Uppsala biochemist Olof Hammarsten, who was no friend of the Allies, would in normal circumstances have backed Haber for the prize. But, with the war still in progress, he invoked expediency in arguing that no award should be made for 1918. Despite opposition from the rest of his committee, the academy took Hammarsten's side, and the delicate matter was shelved.

By the spring of 1919, it was clear that further postponement was impossible, and in November 1919 the academy duly announced the prize for Haber, a decision that had the full approval of Hammarsten, now free to express his true sympathies. The Allies, especially the French, were outraged at the honoring of the man whose research on chlorine and other poison gases had laid the foundations of the age of chemical warfare. In Sweden, too, there was protest. Knowing observers there were quick to interpret the choice of Haber as a victory for a right-wing faction in the Swedish Academy of Sciences. Hjalmar Branting, the leader of Sweden's Social Democratic Party and future winner of a Nobel Peace Prize (in 1921, for his work in the League of Nations), thought the choice scandalous.[25] The academy, in his view, had betrayed Sweden's commitment to disinterested neutrality and allowed itself to be tarred with the brush of barbarism; the decision, as he argued, conveyed not only the conservative elite's pro-German sentiments but also and more seriously its suspicion of parliamentary democracy.

While it is easy to see why Branting took the critical view he did, it could also be argued (and it was argued) that to exclude Haber from consideration would have been precisely to bow to the vagaries of national interest, something that Nobel would have abhorred. That argument, though, carried little weight with the Allies, and the repercussions of Allied indignation soon became public at a special Nobel Prize ceremony in June 1920. All the winners since war had interrupted the ceremonies were invited to what the Nobel organizers presented as an act of reconciliation. But the event turned out very differently. While the German recipients from these years—including von Laue, Willstätter, Haber, Planck (the recipient of the physics prize for 1918, awarded retrospectively in 1919), and Johannes Stark (the winner for 1919)—were all present, there were significant absences among those invited from the Allied countries. Richards, invited to receive his 1914 prize for chemistry, and William and Lawrence Bragg, the joint recipients for physics in 1915, found reasons not to attend. Of the non-German winners in the sciences, in fact, only the British physicist Charles Barkla was present.[26]

Barkla's presence failed to ease the embarrassment. He duly delivered the delayed Nobel lecture that he would have given, but for the war, as the winner of the physics prize in 1917. In rehearsing the customary niceties of a presentation of his own work (on the secondary radiation from metals exposed to X-rays), he did what was expected of a Nobel laureate. But in a separate speech at the Nobel banquet, he also made a point of complimenting the Swedish Academy of Sciences on its disregard of nationality and of singling out two recent German recipients of the physics prize for special praise: Max Planck and, rather more fulsomely, the resolute German nationalist and future champion of the Nazi cause Johannes Stark.[27] Understandably, what Barkla may have intended as a way of distancing himself from the animosities of the war did nothing for his reputation in Britain. There he never shook off the discredit that flowed from his idiosyncratically resolute interpretation of the J phenomenon, a "scientific error" that he pursued for the rest of his henceforth somewhat marginal career as professor of natural philosophy at Edinburgh.[28]

The new Allied order

As was evident in the postwar friction surrounding the Nobel Prizes, the Allies were in no mood for forgiveness, despite their own less than glorious moral record. They were resolved to restructure international science in ways that would ensure their own dominant position and marginalize their former enemies. Despite the plentiful internationalist rhetoric that cloaked them, their intentions were unequivocally vindictive, and they left their mark for many years to come. The process began with three conferences of delegates from the national scientific academies of the Allied countries in London, Paris, and Brussels between October 1918 and July 1919. At these, a new world order for science was established, in which the Central powers would have no place. The main vehicle for the change was the International Research Council (IRC). Formally created in Brussels on July 28, 1919, at the end of the third of the inter-allied conferences, the IRC was successor to the now-disbanded International Association of Academies.[29] While it broadly inherited the IAA's functions, however, its disciplinary and geographical brief was narrower. One change was that the humanities and social sciences would henceforth have their own separate organization, the Union académique internationale. The other, far more controversial, change was that only "adhering bodies" representing the Allied countries and, by invitation, the former neutral nations would be allowed to join the IRC.[30] One consequence was that the IRC began as a body of only sixteen founding members, with little claim to be representing world science.[31]

Politically significant though the IRC was, the intellectual consequences of its policies were felt most keenly within the individual disciplinary unions whose foundation it encouraged from its first meeting. Since the new bodies were subject to the IRC's regulations, it followed that scientists from Germany and its former allies had no right to participate in the work of the unions, particularly and most damagingly in the congresses for which the unions were now responsible. In the same spirit, it was even determined that, at least until the statutes were reviewed at the end of 1931, French and English were to be the sole recognized languages of the IRC, and the use of German would be forbidden at international gatherings under both its and individual unions' jurisdictions. [32]

In the first flush of victory, there was something understandably satisfying about such a boycott. But many in the Allied countries were only too aware that they were willfully turning their backs on a formidable scientific tradition. The evidence that the punitive strategy risked doing more to harm than to favor Allied interests was overwhelming. In chemistry in a typical prewar year, 1909, 45 percent of the items cited in *Chemical Abstracts* were from German publications and in German, compared with proportions of 20.1 percent, 13.4 percent, and 13.2 percent from American, British and Commonwealth, and French journals, respectively.[33]

Physicists, for their part, had only to look at the consequences of the boycott for the prestigious Solvay conferences in their discipline, which chose to follow the IRC's policy on exclusion. At the first two of these conferences, held in Brussels in 1911 and 1913, physicists from Germany had been among the leading participants: Nernst, Wien, Einstein, Planck, von Laue, Arnold Sommerfeld, Emil Warburg, Heinrich Rubens, and Eduard Grüneisen had attended at least one and in several cases both of the conferences. But no Germans were invited to the first postwar conference in 1921, except for Einstein, who had not signed the Manifesto of the Ninety-Three but who declined the invitation in any case in protest against the conference's policy with regard to the Central powers. The boycott was extended to the 1924 conference (with an exception being made for Erwin Schrödinger, Austrian but now established in Zurich) and was only abandoned for the momentous fifth conference of 1927 in which Einstein, Planck, Schrödinger, Heisenberg, and Max Born all participated, along with Niels Bohr and others, in debating the foundations of quantum theory around the conference theme of "Electrons and photons." In the Solvay conferences on chemistry, the pattern of absence and eventual reintegration was similar. Germans were not invited to the first and second conferences in 1922 and 1925, but they returned in force at the third conference in 1928.

The blanket ostracism of the scientists of the Central powers jeopardized personal friendships and collaborations, even if informal channels allowed many of these to survive well enough. It also provoked responses ranging from righteous indignation on the part of the victims of the measures to agonizing remorse at a course of events that had allowed disagreements between nations to degenerate into warfare, with its consequence in victors' justice. Entrenched at one end of the moral spectrum was Wilhelm Wien, who (like Haber) saw no reason why a man of science who had served his country in war should apologize for his actions. The response of the physicist Max Planck, John Heilbron's "upright man," was more nuanced, and anguished.[34] Planck had welcomed Germany's participation in the war as a defensive move that would give voice to German honor and unity, and it was in that elevated spirit that he had signed the Manifesto of the Ninety-Three, though apparently (like Emil Fischer) without seeing the text beforehand.[35] By early 1915, however, Planck was troubled by the consequences of a document that he believed might well breed animosity for years to come, whatever the outcome of the war. In the following year, in an open letter to the physicist Hendrik Lorentz, whose Dutch nationality set him outside the conflict, Planck elaborated his core convictions. These were, in his words, that "there are realms of the spiritual and moral world which lie beyond the strife of peoples" and that an engagement in these realms and continued personal respect for the subjects of an enemy country were "in no way inconsistent with warm devotion and strenuous work for one's own Fatherland."[36]

It was in this spirit of compromise that Planck did what he could to temper the aggressive nationalism of the Alldeutscher Verband (Pan-German League), an ultranationalist organization founded in 1891, to which a number of his academic colleagues belonged. Through that action, as in his personal correspondence and his successful campaign to prevent the Prussian Academy of Sciences from excluding members from enemy nations, he made plain his regret at having signed the manifesto. Yet even the conciliatory Planck would not go so far as to make a public retraction, despite the urging of Einstein and Lorentz that he should do so.[37] After 1918, the dignified moderation of his patriotism, allied to the respect and widespread affection in which he had been held before the war, was to make his personal reintegration in international science a relatively easy matter, as it was later to allow him to survive the Nazi years while retaining his position as an honored representative of German physics in the world community. But, for many others, there was no way back. In sharp contrast with Planck, Wilhelm Ostwald was among those whose life was profoundly affected. His elimination from the International Committee on Atomic Weights in 1916 (after ten years of service) and his loss of influence in internationalist causes, notably the promotion of Ido, hurt him deeply. Equally wounding, though less immediately so, was the Allies' appropriation of his plan for a world center for chemistry and its transformation into the project that eventually took shape in 1927 as the Maison de la Chimie in Paris.[38]

The gratuitous marginalization of Ostwald was shameful enough. But it was eclipsed in its cruelty by the humiliating treatment of Wilhelm Förster, the long-serving former director of the Berlin Observatory and an embodiment of prewar cosmopolitanism, both in his science and as president of the Association scientifique internationale espérantiste, a society founded in 1907 to promote the use of Esperanto for scientific communication.[39] Förster's skill in precise measurement had drawn him to the work of the Bureau international des poids et mesures, and he had been a natural choice as a founding member of the International Committee for Weights and Measures, established to implement and oversee the terms of the Convention on the Metre in 1875; in that capacity, he had done more than anyone to bring Germany into the metric system. As a key figure from the start and president of the committee from 1891, he had come to embody the internationalist spirit of the BIPM and was anxious to see its activities resume as quickly as possible after the war. But when he arrived in Paris for the committee's first postwar meeting in September 1920, the other members refused to sit with him.[40] Förster (like Ostwald) had committed the sin of signing the Manifesto of the Ninety-Three, and had done nothing to distance himself from its sentiments. He now had no choice but to resign his presidency. It was a sorry episode, endured by a frail, respected figure in his eighty-eighth year. As it happened, Förster's exclusion from the committee was also illegal, since the Convention on the Metre was among a small number of treaties and other agreements that had not been subject to the exclusions specified in the Treaty of Versailles.[41]

Painful though many individual experiences were, they were less damaging in the long run than the policies enacted by the International Research Council and the disciplinary communities under its control. The gravest harm was done by the requirement that unions affiliated to the IRC should follow the parent body's restriction of membership to the former Allies and neutral nations. The creation of the unions began in an atmosphere of unchallenged acquiescence in the IRC's principles. The Central powers were duly excluded from the four unions that were set up in July 1919 at the same conference in Brussels at which the IRC itself was founded. All four unions—for radio science (Union radio-scientifique internationale, URSI), geodesy and geophysics (International Union of Geodesy and Geophysics, IUGG), astronomy (International Astronomical Union, IAU), and chemistry (International Union of Pure and Applied Chemistry, IUPAC)—had antecedents in which Germany had been a leading participant. Of the four, URSI (the abbreviation normally used, even in the Anglophone communities) had the shallowest, though still unequivocally international, prewar roots. These lay in the Commission internationale de télégraphie sans fil, created in 1913 to coordinate work in radio telegraphy.[42] The other three drew on far longer histories of international cooperation, with in all three cases a strong German presence that had continued right up to the war.

In geodesy, a tradition of coordinated observing across national borders had sprung from initiatives in German-speaking central Europe that went back to the early nineteenth century. At that time, Friedrich Bessel and Carl Friedrich Gauss had made important pioneering contributions, and in 1862 a Prussian military officer and pupil of Bessel, Johann Jacob Baeyer, had been instrumental in securing the agreement of sixteen countries (seven of them German states) to collaborate in geodesic measurements.[43] Two years later, a first international geodetic conference, in Berlin, had established two powerful bodies—a "Permanent

Commission" with scientific responsibility and an executive "Central Bureau"—to oversee the work of what was formalized soon afterward as the International Geodetic Association (IGA). From its headquarters in Potsdam, the IGA went on to play a central role in triennial congresses and shared research programs that continued right up to the war. After such a history of German-led cooperation, it was hard to see the IUGG, now with no German presence, as anything but an eviscerated successor to a conspicuously effective prewar structure, and for some years the loss weighed heavily on research in the field.

Astronomy likewise had been a successful multinational pursuit in the nineteenth century: the Carte du ciel project of 1887 and the International Union for Co-operation in Solar Research, established in 1904, had led the way in promoting ventures that brought together astronomers from across the world. Many of these ventures, though, grew from German initiatives. From its foundation in 1863, at least until the First World War, the German Astronomische Gesellschaft functioned de facto as an international society.[44] It had a significant proportion of foreign members, and although its meetings were normally conducted in German, it was ready to publish contributions in any language. The society's internationalist credentials were exemplified in its Zonen-Unternehmen, a project for the systematic observation and cataloguing of the northern stars classed as the Bonner Durchmusterung (the Bonn group) and originally defined as being observable with the telescope of the Royal Observatory at Bonn. The project, begun shortly after the founding of the society and essentially complete by 1914, rested on a complex division of labor among twenty observatories, each of which undertook to observe a particular zone of declination. While the project's German tone and inspiration were beyond question, the fact that thirteen of the observatories were located outside Germany marked it as a genuinely international enterprise.[45]

The new union for chemistry, IUPAC, had a record of prewar cooperation comparable with that of the IAU. Although its immediate predecessor, the International Association of Chemical Societies, had been founded as recently as 1911, chemists had been among the pioneers of scientific congresses from the time of the first international chemical congress, in Karlsruhe in 1860. Thereafter, working parties had met during and between congresses to discuss nomenclature and other conventions of chemistry. The discussions were never easy, and it was a major achievement when basic agreement on the nomenclature of organic chemistry was achieved at a specially convened international conference in Geneva in 1892.[46] Over four days at the conference, thirty-four representatives from nine countries, under the chairmanship of the Alsatian chemist Charles Friedel, laid the foundations of a system of "official names" for organic compounds that would convey their chemical structure. As discussions during the 1890s and the early years of the new century were to show, the Geneva Nomenclature left many questions unresolved, and it never achieved universal assent; divergences between the approaches of the French (as represented by Friedel) and the Germans (represented by Adolf von Baeyer) proved especially hard to bridge. The fact remained, however, that a precedent for successful collaboration in pursuit of an internationally approved system had been established. It was a precedent that gave the IACS something substantial to build on when, in turn, it took up the problem of organic nomenclature during its brief existence before the war.[47]

In all the sciences, the effects of the war on strong traditions of cooperative endeavor caused soul-searching. It also raised tensions between the various national groups involved in setting up the new postwar disciplinary structures. Broadly, representatives from France and Belgium reflected the particular wartime hardships of their countries by taking a firm stand on the exclusion of the Central powers from the IRC and the disciplinary unions; they wanted the exclusion not only to be strictly enforced but also to remain in place for a significant length of time, even indefinitely. Initially, their colleagues from Britain and the United States went along with this unforgiving line, out of solidarity with French and Belgian sensibilities and in the punitive spirit of the Treaty of Versailles. But there were always dissenting voices. Among the individuals who spoke up against the boycott from the start were the British mathematician G. H. Hardy and, in Sweden, Mittag-

Leffler and Arrhenius.[48] A sharp divergence of opinions also separated the two Japanese delegates to the conferences and other meetings of 1918–1919 that led to the founding of the IRC. While the chemist and main instigator of the plans for a Japanese National Research Council, Sakurai Jōji, supported the IRC's position, Tanakadate Aikitu, a physicist and colleague of Sakurai at Tokyo University, was resolutely opposed.[49] Tanakadate, who had studied for two years in Glasgow with William Thomson in the late 1880s and spent a year at the University of Berlin, had experienced the realities of international collaboration as a contributor to the field of geomagnetism and a regular participant in congresses.[50] In his reservations with regard to the IRC he was by no means alone among Japanese scientists, many of whom valued their traditionally strong links with Germany. Yet political expediency prevailed, and Sakurai's determination that Japan should stand shoulder to shoulder with the other Allied powers carried the day. By the time the creation of the IRC and the first four unions was finally approved in July 1919, the proposals for exclusion went through without opposition.

Lost illusions: A postscript to war

Even as exclusionist policies were carrying all before them in the IRC and the unions, those who had harbored more prophetic, apolitical visions of internationalism returned to their old causes with undiminished vigor as soon as hostilities ceased. In 1918, when the Armistice had not yet been signed (though with the end of the war in sight), Hendrik Christian Andersen published luxurious English and French editions of a substantial supplement to his *World Centre of Communication* of five years earlier.[51] The book, prepared with his late sister-in-law Olivia Cushing Andersen, who had died in 1917, reaffirmed Andersen's vision of a "science of legitimately controlling international affairs." In the optimistic words of his preface, this was to be a "counter-science" able to work against the misuse of science for military purposes and draw the peoples of the world together in pursuit of "human development and harmony."[52] The tone and language suggested that nothing had changed. Indeed, by devoting most of the book to two lengthy essays by authors addressing the practicalities of bringing the World Centre to fruition, Andersen could claim that support for his cause was growing. In one of the essays, the Milanese jurist and philosopher Umano summarized the principles of his monumental *Positive Science of Government*.[53] In the other, the former New York University professor Jeremiah Jenks treated the economic advantages of the World Centre idea. For Umano and Jenks, the plan's feasibility was not in question. Nor was it for Andersen, who resumed his campaign for support in influential circles. To a formal letter accompanying the copy of the supplement that he sent to Lord Curzon, he added a handwritten request that, in his capacity as chancellor of the University of Oxford, Curzon should give the project whatever publicity he could.[54] Although Curzon was impressed by Andersen's scheme, he seems to have taken no action. Nor, so far as we know, did other eminent recipients, such as the French president Raymond Poincaré, or the future president of the United States, Herbert Hoover, admired by Andersen for his humanitarian relief efforts in wartime Belgium.[55]

Like Andersen, Otlet too resumed his mission. He did so in the same spirit of renewed optimism, despite the loss of his son Jean, who had been killed in the early weeks of the war (although Otlet only learned of his death in 1918).[56] An immediate concern was the Institut international de bibliographie, which he and La Fontaine found in a surprisingly good state, despite four years of greatly diminished activity. Associated enterprises too had to be brought back to life, under the overarching control of the resurrected Union des associations internationales. A meeting with the Belgian prime minister Léon Delacroix, at the head of a government of national unity embracing Catholic, liberal, and socialist interests, was encouraging. Delacroix saw to the reinstatement of the prewar governmental budget and arranged for an improved installation for the museum and the IIB and its library in some fifty rooms in the wing of the Palais du cinquantenaire where the museum had been housed since before the war.[57] The adoption of the title Palais mondial for their various enterprises conveys the confidence that Otlet and La

7. Paul Otlet (center) and Henri La Fontaine (right, next to his wife) outside the
Palais du cinquantenaire, Brussels, c. 1930. Shortly after the First World War, Otlet
and La Fontaine gave the name Palais mondial to the premises they occupied in
the Palais du cinquantenaire. The Palais mondial, commonly referred to as the
Mundaneum from 1924, housed the Institut international de bibliographie, Musée
international, and offices of the Union des associations internationales, and was
the setting for sessions of the Université internationale in the early 1920s and 1927.
Courtesy of the Mundaneum, Mons.

Fontaine felt, and the effect was immediate (see figure 7). In April 1921, on the occasion of a fair in the Parc du cinquantenaire, a collection that had attracted just under thirteen thousand visitors a year before the war welcomed two thousand a day, with a peak on one day of eight thousand.[58] And for a while the popularity continued; by 1923, according to Otlet, the number of visitors for the year reached fifty thousand.[59]

Other new departures too lent encouragement. In December 1919, after months of lobbying for recognition by the powers involved in laying the foundations of the League of Nations, the UAI announced the most ambitious of all its ventures, an international university that would offer its first courses in the following September, also in the Palais du cinquantenaire.[60] The plan, with roots in proposals for closer transnational cooperation between universities, professors, and students going back to before the turn of the century, was for an institution in which an elite of 1 percent of the world's students would supplement whatever program they had followed in their own university with an additional year of international study.[61] The year was to be divided between six months at the headquarters of the Université internationale, where students would be taught by professors seconded from the world's leading universities, and six months of travel on a "tour du monde universitaire." In keeping with the university's insistently international tone, courses would be designed to promote a global "mentality" refined through a comparative perspective on the "great questions" of world culture. The rhetoric was grandiose and confident. But no one could miss how uncomfortably it sat with the all-too-predictable hallmarks of undimmed postwar recrimination. Teachers and students from Germany and the other Central powers were excluded, and French and English were imposed as the official languages, with professors using other languages only if they could not do otherwise.

Despite the selective conception of internationalism that belied its name, the Université internationale got off to a bright start. In the inaugural trial session, which took place from September 5 to September 20, 1920, during a fortnight of lectures and meetings that Otlet designated as a first "Quinzaine internationale," 143 classes were given by forty-seven professors from ten countries, with more than a hundred formally enrolled students attending and a similar number present as "auditors."[62] Otlet and La Fontaine set the tone in a formidable list of speakers, including many from outside Belgium. Andersen, Pierre de Coubertin, and the French educational administrator and leading positivist Emile Corra, along with Patrick Abercrombie, professor of civic design at the University of Liverpool and pioneer of the British "New Towns" movement, and Edgar Milhaud, peace activist, champion of the League of Nations, and professor of political economy at the University of Geneva, were all well-known figures whose internationalist credentials were not in doubt. Encouraged by the initial success and a cautious show of interest by the League of Nations, further sessions were held in 1921 (with sixty-nine professors giving 174 hours of teaching)[63] and 1922 (ninety-six classes taught by seventy-two professors).[64] But the tensions surrounding the French and Belgian occupation of the Ruhr coalfields in pursuit of unpaid German reparation payments in January 1923 presaged new difficulties and made the abandonment of that year's session all too predictable, despite evidence of some last-minute planning.[65] And worse was to come. Apart from a stray, much-reduced session in 1927, the university died, a victim of the worsening international situation and, as the economic crisis deepened, for want of funding.

The pattern of initial postwar optimism giving way to a loss of momentum was evident in all of Otlet and La Fontaine's initiatives. The UAI's journal, *La Vie internationale*, revived in November 1921, managed only one issue before ceasing publication. And still darker clouds gathered as the consequences of Delacroix's departure as prime minister and his replacement, in December 1921, by the far less sympathetic Georges Theunis made their mark. Thereafter, governmental attitudes fluctuated between indifference and outright hostility. The government's predatory intentions with regard to the space in the Palais du cinquantenaire devoted to the UAI and its associated activities presented a particular threat. In 1924, a history of sustained minor harassment came to a head with the installation of a trade fair in the Palais mondial's premises and the removal of much of the UAIs' furniture and the museum collection into storage.[66]

The decision to move against the Palais mondial and the UAI, though not unexpected, had an element of intense irony in that the fair was to be devoted to rubber, the product of the Congo Free State through which King Leopold II had enriched himself from the 1890s until he finally ceded the territory to the Belgian government in 1908. Colonial exploitation of such a ruthless kind, with the cruelties that it engendered, was among the denials of human dignity against which Otlet and La Fontaine had protested most vehemently. Their making the Palais mondial available for a three-day Pan-African Congress as part of the "Quinzaine internationale" of 1921 had already made plain their commitment to the anticolonialist cause.[67] It was a characteristically principled gesture, but one that distanced them from a broad swath of Belgian public opinion. The consequences, not only for the episode of the rubber fair but also for relations with the succession of conservative administrations that governed Belgium for most of the 1920s, were damaging.

A partial retreat by the Theunis government, in the form of a decision to allow the UAI to resume its occupancy of the Palais du cinquantenaire after the rubber exhibition ended, betokened a measure of contrition; by 1926, half of the rooms of the museum had been refitted, and in the following year work began on the reinstallation of the library. But nothing could counter perceptions that the UAI's various causes had entered a phase of terminal, if slow, decline. As La Fontaine became more involved in his work as a university professor and senator, and with the League of Nations, Otlet was left an increasingly isolated and embittered figure. Disappointment, though, did not betoken capitulation. Under the collective name Mundaneum, which Otlet adopted for his Brussels-based activities in 1924,[68] what remained of the Palais mondial continued its impecunious existence. The bibliographical institute even struggled on into the early 1930s, though as a shadow of what it had once been. By then, bibliography in the sciences had become firmly entrenched as the preserve of experts in the various disciplines. Finely focused and better-resourced initiatives such as *Chemical Abstracts* and *Biological Abstracts,* a relative newcomer but with a similar brief,[69] served the needs of scientists more effectively than Otlet's seriously underfunded program of universal bibliography could ever hope to do. Moreover, what remained of the program slipped inexorably from his control. Responsibility for the Universal Decimal Classification, in particular, passed to his younger Dutch collaborator, Frits Donker Duyvis, who acted as secretary to the committee for the updating of the UDC from 1921. The loosening of the reins entailed significant compromise, in particular over the priority that Otlet and La Fontaine had always given to the maintenance of a single central repository, which Duyvis never accepted.[70] Otlet's eventually strained relations with Duyvis are not hard to understand.

As his early postwar optimism subsided, Otlet saw challenges on every side. For the UAI as a whole, the most direct challenge came from the International Committee on Intellectual Co-operation (ICIC), a creation of the League of Nations of which, from its foundation in 1922, Otlet was always suspicious and not a little jealous. The structure of the ICIC, as a committee of a dozen or more internationally minded public figures, was quite different from that of the UAI. But its aims, as the league's cultural wing with responsibility for promoting international exchanges and partnerships between scientists and scholars, overlapped uncomfortably with those of the UAI. After a brief attempt at collaboration, Otlet attacked the ICIC publicly, accusing it of usurping the functions of the UAI and doing so all too successfully.[71] His position of weakness was evident. The UAI could not match the eminence and commitment of the ICIC's leading members; these included the Oxford classical scholar Gilbert Murray, the French philosopher Henri Bergson, the Spanish writer and pacifist Salvador de Madariaga, and, among scientists, Marie Curie, Albert Einstein, Hendrik Lorentz, Robert Millikan, Paul Painlevé, Tanakadate Aikitu, and Torres Quevedo.[72] Nor, by an even larger margin, could it match the effectiveness of the ICIC's Paris-based executive and publishing wing, the Institut international de coopération intellectuelle, established in 1926.[73]

La Fontaine lived on until 1943, Otlet until 1944, still campaigning, still speaking, and drawing what comfort they could from the remaining few who took their ideas seriously. Otlet's ideas had a distant legacy in the work of the logical positivist and leading member of the Vienna Circle Otto Neurath, whose admiration for Otlet's work informed his method for conveying information through an "international picture language" based on graphic symbols, or "Isotypes."[74] Neurath was still working on his isotypes and hence perpetuating the memory of Otlet at the time of his death, in England, in 1945. But the dream that was embodied in the various activities of the UAI and the Mundaneum effectively ended in 1934, when the Belgian government expelled the Palais mondial from the Palais du cinquantenaire to provide more space for the collection of the Royal Museums, which it continues to house today.[75] To Otlet's grief, the contents of the UAI's offices, the museum, the library, and the IIB's cabinets, with their more than fifteen million cards, were moved to premises that made meaningful access impossible. The episode had a special poignancy in that the expulsion of the Palais mondial happened in the year in which Otlet's masterly manual of bibliography and library science appeared.[76] But even the landmark *Traité de documentation*, in a brief "Postface" to which Otlet voiced a sad final protest, could not conceal the fact that the higher aspirations that lay behind it were in tatters.[77] With Otlet continuing to work as best he could from home, the collections eventually found their way into buildings elsewhere in Brussels, in the Leopold Park. There they moldered from 1941 until their discovery in 1972 by Otlet's first biographer, W. Boyd Rayward, in total disorder and remembered only by a dwindling band of "Amis du Palais Mondial."[78]

Andersen became similarly disenchanted. His gathering disillusionment owed much to his scepticism with regard to the League of Nations ("this diplomatic and political international body of retired politicians"[79]) and his indignation at the way, as he saw it, Otlet and Le Corbusier had appropriated his ideas in the unrealized plan for a world city of their own in Geneva.[80] Yet as late as 1930, he was still ready to send out a copy of the *World Centre of Communication* to the British trade unionist and Labour cabinet minister Arthur Henderson, a man he admired for his commitment to the cause of peace and the League of Nations, especially during his two years as foreign minister in Ramsay MacDonald's government between 1929 and 1931.[81] In the mid-1930s Andersen briefly returned to the fray once again, when he launched a new appeal for support through the international society, "World Conscience," that he had founded before the war. The supporting publicity literature, undated but almost certainly put out in 1934, is notable for its mention of a warm endorsement of the World Centre project by Benito Mussolini, who had received and encouraged Andersen in 1926 and even offered a site for the city between the ancient port of Ostia and Fregene, about fifteen miles from Rome.[82] Despite the brevity of the interview, which appears to have lasted for no more than ten minutes, Mussolini's interest had left its mark. As Andersen wrote (eight years after the event), "The fact that the scheme is approved by so advanced a thinker and statesman as the great Fascist leader should be a weighty argument in its favour."[83] In 1935, in a radio broadcast for listeners in the United States in which he described his universalist ideals and plans for his world city, he again took the opportunity of praising Mussolini, this time as the fount of the spiritual and material regeneration of the Italian people.[84] But by now his project aroused little interest, and until his death in 1940, Andersen's own priority was his art.

The collapse of the optimism and belief in the possibilities of human progress that had inspired so many reforming spirits at the beginning of the century betokened the effects of a war that, for scientists and other intellectuals, lingered on for several years after the Treaty of Versailles in 1919. It also reflected the passage of leadership in international cultural activities from the individuals and essentially voluntary organizations of the kind that the UAI represented to governmental and other officially recognized bodies of far greater weight and efficiency.

The reordering of the world of learning that began immediately after the Armistice had given institutional expression to these changes. In the euphoria of victory, the Allied scientific communities had fashioned a structure dominated by the leading victorious nations and weakened by divisions that perpetuated the animosities of the war until the mid-

1920s. True internationalists who saw openness and universal access to knowledge as higher priorities than political advantage and retribution for the supposed crimes of war saw much to alarm them in the new order. Quite apart from the ostracism of the Central powers, the postwar conception of world science had obvious flaws. One was that the "world" of the International Research Council and its unions did not embrace a number of countries, in addition to the Central powers, that stood apart by choice or necessity. Among these was Russia, before 1914 a major scientific power where science of any kind was now having to struggle to survive the succession of revolution, civil war, and the eventual establishment of the Soviet Union in 1922. Beyond Europe and the American continent, the absentees were even more apparent, notably in the East, where only Japan occupied a significant place during the IRC's early years.

It was not until the later 1920s that committed internationalists could draw some satisfaction from the signs of a return to normal relations between former enemies and an opening to new settings for scientific activity, notably in Latin America, where Venezuela, Colombia, and Brazil were among the leaders of an assertive scientific movement closely bound up with rising economic prosperity and a sense of national identity. But even as old animosities subsided and hopes of a more cooperative spirit revived, changes in the political and cultural climate, in which science and national interests came together in a quite new symbiosis, delivered an unambiguous reminder that the war and its immediate aftermath constituted an unbridgeable watershed between past and present. Internationalists who had known the prewar world might look back nostalgically to a time when a dominant scientific cosmopolitanism had sustained a web of interactions that took little account of national boundaries and the particularities of language and cultural tradition. But they knew that rebuilding that world in anything resembling its old form was by now an unrealistic option.

CHAPTER THREE

The Legacy of a Fractured World

From the mid-1920s, the disparate feelings of unease about the boycott of the former Central powers began to merge into a concerted move to reestablish normal relations within the world of science. Until then, the victors, especially France and Belgium, perceived themselves as firmly in the ascendant. While their opinions on the wisdom of the International Research Council's policy varied, they were agreed that if and when the policy was abandoned, the process would have to be handled carefully. It was seen as especially important that there should be no taint of capitulation to external calls of the kind being voiced by the neutral nations in favor of rapid, unconditional normalization. The nations targeted by the boycott, for their part, were not cowed by their exclusion; within two or three years of the Armistice, they were adapting positively to the sanctions that the IRC and other international bodies had imposed on them. Germany in particular showed every sign of reemerging as a major scientific power well able to make its way in a world of new alliances, extending even to the Soviet Union.

One conclusion to be drawn from the postwar readjustments was that, as a motor of policy, the idealistic all-embracing internationalism that had fired so many ventures before 1914 had been seriously undermined. Scientists on both sides had served their countries to good effect. They were proud to have done so, and their various peoples were grateful to them. An immediate consequence was that science was set to rise in the priorities of governments virtually everywhere and so to become ever more closely bound into the machinery of state. Recognition of the material benefits of scientific and technological research as a force for economic as well as military success was essential to this heightened awareness of what scientists and engineers could contribute to national well-being. The cultural distinction that came with a high profile in science had its value as well. A Nobel Prize, an invitation to host an international congress, or a great scientific achievement had certainly mattered before the war. But now, in a process that I associate with a "national turn" in science, beginning in the 1920s and accelerating through the 1930s, the prestige that such successes bestowed mattered more than ever.

The road to normalization

The measures agreed upon at the third inter-allied conference of academies in Brussels in July 1919 had been conceived with all too little thought for the consequences for science and scholarship. A world of learning that had no place, or only a peripheral place, for the former Central powers was a seriously diminished world. Moreover, it was diminished for everyone concerned. As more temperate reflection succeeded the initial wave of the victors' triumphalist rhetoric, a sense of the damaging consequences for Allied interests and the futility of the boycott gained ground. The change of heart was fed by evidence that even in the years of resolute formal exclusion in the early 1920s, the Allies' hard line was never entirely effective. It did little to curb individual contacts, which continued through correspondence and visits.

Aid packages too, in particular from the United States, routinely crossed the divide that the Allies were committed, in principle, to sustaining. Aimed at supporting science and medicine in parts of Europe most affected by unfavorable exchange rates and other adverse conditions, they included former enemy nations among their beneficiaries. In this charitable work, the Rockefeller Foundation played an exemplary role through grants that did much to spread the knowledge of American practices in medical education, especially in public health.[1] The new nations of Czechoslovakia, then Poland, were initially favored, and after some hesitation, Hungary and Bulgaria too were judged stable enough to receive help. By 1922, early Rockefeller grants to medical schools in Vienna, Innsbruck, Graz, and Budapest were being supplemented by support for German laboratories and gifts of journals to medical schools.[2] In related areas, research in human biology and biochemistry too was consistently a Rockefeller priority. And further aid was to follow, as the foundation's Division of Medical Education instituted a generous allocation of resident fellowships for physicians and medical scientists in Germany and elsewhere in central and eastern Europe who wished to study abroad.[3]

At the same time as they were receiving such help, German scientists were showing conspicuous resourcefulness in helping themselves. With many institutional avenues denied them by the decisions of 1919, this required ingenuity. But scientific communities in the neutral countries and former wartime allies to the east harbored valuable sympathizers. So too did the Soviet Union, once (as prerevolutionary Russia) in the Allied camp but, from April 1922, committed to a regularization of diplomatic relations with Germany following the German-Soviet Treaty of Rapallo. In turning to the Soviet Union, which now and for many years to come remained outside the IRC and its successor, ICSU, Germany's scientists were resuming contact with colleagues traditionally open to German ways of doing science. A number of strong alliances were soon cemented, often on the basis of personal friendships between colleagues working outside the mainstream world of international congresses and other meetings.[4] The benefits were evident not only in joint ventures but also, in some measure, in the continued use of the German language. The launch of a new bilingual German-Russian medical journal marked the eastward turn, as did medical reading rooms, supplied with German periodicals and books, in Moscow and St. Petersburg, a German-run bacteriological station in Moscow, and a number of other German-Soviet collaborations and conferences.[5]

At first, scientists in the Allied countries took no particular notice of the rapprochement. Soon, however, they could not ignore the disquieting implications of Germany's inexorable return toward the status of a leader in world science and the challenge that this implied for the disciplinary unions. In the International Union of Geodesy and Geophysics, the implications were aggravated by a history that had begun during the war, when a group of neutral nations had formed an "Association géodésique entre les états neutres" to continue collaborative research independently of the belligerents and maintain contacts with the headquarters of the International Geodetic Association in Germany's distinguished Geodetic Institute in Potsdam. When the IUGG was formed in 1919, the Reduced Geodetic Association (as the Association géodésique was commonly called) saw no reason why it should immediately abandon its conspicuously successful program to make way for the new body and so, by implication, endorse the IRC's policies. The result was a damaging rivalry between the former members of the Reduced Association and the IUGG that endured into the later 1920s.[6]

Other unions were affected by tensions of a different kind that weighed ever more heavily as new nations joined. The International Astronomical Union, for example, had been formed in 1919 with a core of seven member countries, all Allies: Belgium, Canada, France, Great Britain, Greece, Japan, and the United States. By 1922, however, nineteen countries were represented at the IAU's General Assembly in Rome, and three years later, at the General Assembly in Cambridge, the number had grown to twenty-two.[7] In chemistry, IUPAC underwent even more spectacular growth, from a close-knit group of five founding members (Belgium, France, Great Britain, Italy, and the United States) to a total membership of twenty-eight by 1925.[8] Expansion on such a scale left the former Allies in a small minority, with the dilemma of coping with a boycott that, in many eyes, had become a damaging anachronism.

The first cracks in the IRC's public façade of solidarity appeared as conference organizers wrestled with the intellectual consequences of the boycott for their particular activity. As early as June 1923, the organizing committee of the international conference on phytopathology and economic entomology in Wageningen in the Netherlands did not hesitate to invite German, Austrian, and Hungarian colleagues, along with representatives of all the Allied powers.[9] And two years later the growing conviction that the IRC's selective notion of internationalism was doing a disservice to science provoked an even more flagrant breaking of ranks at the Third International Congress of Entomology in Zurich.

The Zurich congress was the first in the subject since the predominantly (though by no means exclusively) Anglophone Oxford congress of 1912, the congress planned for Vienna in 1915 having fallen victim to the war. The rifts in the international community with regard to the IRC were laid bare in the decision to conduct and publish the proceedings almost entirely in German.[10] Significantly, among twenty national delegations there was none from Belgium, and the French were represented by a single provincial society, the Société normande d'entomologie from Caen.[11] Of the individuals present, a strong British contingent, anxious to resume contact with German-speaking colleagues regardless of the IRC's interdiction, contrasted sharply with the solitary person attending from France. The brisk assertion, in the unsigned preface to the conference proceedings, that it was time to emerge from the "Kriegspsychose" that had afflicted science since the war conveyed the Swiss organizers' frustration at continued French and Belgian resistance.[12] So too, and even more pointedly, did their regret that Pierre Lesne, a former president of the Société entomologique de France, had felt obliged to withdraw his initial support for the congress, apparently under the pressure of opinion in the French scientific community. But the absences did nothing to diminish the sense of satisfaction, expressed in the opening addresses, that the "scientific solidarity between nations" had been triumphantly affirmed.[13]

While the earliest calls for the abandonment of the boycott came from the former neutral nations, by the mid-1920s American delegations too were emerging as leading advocates of normalization with a minimum of delay. In the IAU, smouldering American discontent surfaced at the Cambridge General Assembly of 1925 in a forthright statement by the lunar astronomer E. W. Brown of Yale in support of a number of requests for the union to open its doors to all countries, including Germany and its former allies.[14] If the IRC's restrictive statutes were not modified, Brown warned, the US delegation would be unable to recommend American participation in future congresses. Within IUPAC, a corresponding

American voice was that of William A. Noyes, a senior professor at the University of Illinois at Urbana-Champaign, former president of the American Chemical Society, and a determinedly international figure who had worked hard to allay Franco-German animosities.[15] As well as hoping to draw a line under the enmity that had continued to divide chemists since the war, by the mid-1920s Noyes faced the immediate specter that the union's impending review of organic nomenclature would lack any semblance of international authority, unless German and Austrian chemists were involved.[16]

Advocates as influential as Brown and Noyes could not be ignored. The cause, though, was not easily won, as the US delegation to the Toronto Congress of the International Mathematical Union in 1924 found. During the IMU's General Assembly, the delegation formally proposed that the IRC should be urged to "consider whether the time is ripe for the removal of restrictions on membership now imposed by the rules of the Council."[17] Endorsement of the American position by Sweden, Denmark, the Netherlands, and Norway, backed by Italy and Britain, was predictable. So too was the absence of French and Belgian support, and what began as a bold bid for change was eventually defused with the briefest of mentions in the minutes and an agreement simply to report the proposal to the IRC.[18] The greatest single obstacle to action was the French mathematician Emile Picard, who at the time and for some years to come was president of both the IMU and the IRC. Shortly before the next IMU, in Bologna in 1928, chairman of the organizing committee and congress president Salvatore Pincherle pleaded with Picard to relent in his insistence that German admission to the union could only be considered once Germany had joined the IRC.[19] As a committed internationalist who had studied in France and Germany as well as in Italy, Pincherle made a forceful case. But Picard and the secretary-general of the union, another French mathematician, Gabriel Koenigs, stood by the rules. It was a mark of the very different climate prevailing by the later 1920s, however, that they did so in vain. The standing ovation that greeted the entry of the large German delegation (with David Hilbert at its head) at the Bologna congress signalled that, at least in mathematics, regardless of the IRC and French objections, the reintegration of Germany was finally a reality.[20]

With the authority of the IRC so palpably in decline, it might have been expected that the Locarno Treaties, which brought Germany back into the community of nations in September 1926, would lead on smoothly to the more general reconciliation that the planners of the Zurich congress and the critics of the IRC's rigid stance hoped for and anticipated. But memories of the earlier inflexibility died hard, and they left many leading figures, especially in the disciplinary unions, sceptical of the IRC's internationalist credentials.[21] In the cautiously conciliatory spirit of Locarno on June 29, 1926, the IRC proffered a gesture, inviting Germany, Austria, Hungary, and Bulgaria to join; at the same time, the ban on the use of German in the work of the council and its associated unions was abandoned. But only Hungary accepted the olive branch.

Germany's rejection of the IRC's invitation created a standoff that was not reversed until after the Second World War.[22] Even the general openness of culture during the Weimar Republic had only a limited effect in reconciling the German scientific community to the Allies' conception of international science, a point that Paul Forman has made with respect to the gifted generation of Weimar's physicists.[23] The IRC, for its part, remained unyielding. This meant that the increasingly frequent presence of German delegates at congresses of the various unions from the later 1920s was strictly improper. The irregularity clearly irritated the council's president, Emile Picard. Addressing the IRC General Assembly in 1928, Picard expressed bewilderment at the fact that, even two years on, Hungary remained the sole representative of the former Central powers in the IRC.[24] He should not, of course, have been surprised. The reticence of Germany, in particular, with regard to an organization that had moved so slowly—grudgingly as many would say—to heal that wound of war was entirely understandable.

Within individual unions, the pattern of reconciliation proceeded but did so unevenly. In IUPAC, a significant lobby in favor of normalization was heartened by Bulgaria's admission to the union in 1928. But it was not until two years later that Germany finally joined, though only after intense negotiations involving two supportive IUPAC presidents, both from neutral countries: Ernst Cohen from the Netherlands (1926–1928) and Einar Biilmann from Denmark (1928–1934).[25] Where the power of personal persuasion was less strong, the process was far trickier. Germany did not join the IUGG, for example, until 1937, and it remained outside some other well-established unions, such as the IAU and the International Union of Pure and Applied Physics, until the early 1950s. It was anticipated that the long-planned review of the IRC and its reconstitution as the International Council of Scientific Unions (ICSU) in 1931 might overcome German suspicion. But the changes, which initiated something of a fresh start by giving the unions enhanced voting power and hence greater influence relative to the national representatives, had limited effect. While they were sufficient to induce Bulgaria to join, they failed to bring in Germany or Austria.[26] In due course the old suspicions would almost certainly have abated. That process, however, was soon to be brutally halted by the external force of Nazi policies that discouraged participation in international ventures and diminished any possibility that Germany (no longer in the League of Nations from January 1933) might experience a change of heart.

The national turn

As internationalist sentiment and the cause of reconciliation struggled to recover their voices through the 1920s, the counterweight of expressions of national pride too became more prominent. This was manifested in science, as in other areas of learned culture, in a new insistence on the distinctiveness, and distinction, of national traditions. An early illustration was the exuberance of the French celebrations of the return of Alsace and the parts of Lorraine that had been lost to Germany after the Franco-Prussian War of 1870. Within the recovered territories, the city of Strasbourg had special significance as a potential shopwindow for

French culture and its standing in the world. And the opportunities of a setting adjacent to the German border were duly seized. The decision to hold the 1920 International Congress of Mathematicians in the city, rather than in Stockholm (which had been chosen as host at the Cambridge congress of 1912), bore all the marks of French lobbying.[27] In Sweden Mittag-Leffler was enraged and never recognized the event as an official international congress, referring to it contemptuously as "une affaire française."[28]

What transpired was indeed very much a French occasion, with the eighty-two-year-old Camille Jordan as a venerable honorary president of the congress; Emile Picard, quite new in his presidency of the IRC, as its president; and Gabriel Koenigs as its secretary-general. In his closing address to the Strasbourg congress Picard performed what he saw as his patriotic duty by delivering a blistering rebuke to "those who have excluded themselves from the concert of civilized nations" and laying explicit claim to the same moral high ground that the Belgian Cardinal Désiré-Joseph Mercier had occupied through his much-quoted principle that certain crimes were of such gravity that to pardon them was tantamount to condoning them.[29] Most of the French mathematicians at the congress (eighty of them, out of a modest overall attendance of about two hundred) would have been pleased with such rhetoric. They would also have warmed to the resolute line taken at the first general assembly of the newly constituted International Mathematical Union, which was held as part of the proceedings.[30] All eleven of the formally accredited national delegations to the assembly duly reaffirmed the IRC's founding principle concerning the exclusion of the Central powers. The individual dissenting voices of Hardy and Mittag-Leffler, neither of whom was present, were simply ignored.[31]

Significant though the 1920 congress was, for French national interests the jewel in the Strasbourg crown was the city's university, the Kaiser-Wilhelms-Universität (KWU), as the university was known from its foundation in 1872 until it was replaced immediately after the war by the French Université de Strasbourg. In barely four decades, material investment and the appointment of a particularly gifted body of professors had transformed the university from a poorly financed group of faculties on the margins of the French system of higher education, before 1870, into a showpiece that bore comparison with the leading institutions in the world.[32] The expenditure on buildings (all of them completely new), instruments, and other fittings for the sciences was massive by any standards, and that, allied to the investment of German national pride in the project, gave the KWU a special place in the spoils of war that came France's way.[33]

One response might have been for the Ministry of Public Instruction to fashion a university that would bring together all that was best in the French and German traditions of academic life. But no such conciliatory venture seems to have been considered. German professors, regardless of their distinction, were summarily dismissed, and the appointment of French replacements began under a provisional administration that set to work within barely a fortnight of the Armistice.[34] With French peremptorily imposed as the sole language of instruction, the university was now a French institution serving French interests. That point was theatrically conveyed in November 1919, at the inaugural ceremony for the first full postwar academic year. In the presence of Raymond Poincaré, in his capacity as president of the Republic, the university's new professors and lecturers (184 of them, twenty more than in the German KWU and more than in any other French university at the time, with the exception of Paris) joined with distinguished visitors from sympathetic nations across the world in turning a routine academic event into a great patriotic occasion.[35]

A city imbued with such meaning lent itself perfectly to the celebration that followed three years later, to honor the most admired of the nation's scientific heroes, Louis Pasteur. The celebration marked the centenary of Pasteur's birth, beginning in December 1922. A substantial new biography,[36] a flood of ephemeral literature and souvenirs, and Pasteur's being chosen as the first Frenchman (since the Emperor Napoleon III) to be portrayed on a stamp, all played their part in generating popular excitement, as did receptions, lectures, and special meetings of societies across France. The climax came in May 1923 (five months after the true centenary) with a week of events formally designated and conducted, with full ministerial backing, as an official "commémoration nationale."[37] After a ceremony in the Sorbonne at which representatives of more than fifty national delegations spoke or presented congratulatory addresses, many of those attending (led by the president of the Republic, now Alexandre Millerand, travelling in a presidential train) moved on to locations associated with Pasteur's life in the Franche-Comté region and, finally and inevitably, Strasbourg. As the city where Pasteur had spent the first five years of his academic career and the seat of a recently established Institut Pasteur,[38] Strasbourg had claims (independently of its having been so recently "delivered" from German control) that made it the natural location for a major exhibition inspired by his work.[39] There was a sad irony in the presentation of the exhibition, on the theme of public health, as "international." No representatives of the former Central powers were present among the fifteen participating foreign nations. Likewise, none attended the inauguration of the monument to Pasteur in front of the main entrance to the university. At a celebration planned as a patriotic "glorification" of Pasteur (Millerand's term[40]) and a statement of France's past scientific glories and a program of national renewal to come, a German presence would have been inappropriate, not to say offensive.

Four years later, in 1927, Paris hosted a week of even grander ceremonies, to mark the centenary of the birth of the organic chemist Marcellin Berthelot. As in 1923, the program began with a glittering academic gathering in the great amphitheater of the Sorbonne, and again lectures, fund-raising events, another commemorative stamp, and an admiring new biography[41] contributed to stimulating public interest in a man who, quite apart from his eminence as a chemist, had been a revered champion of the secular republican left (in power once again in the mid-1920s).[42] The high point this time was the laying of the foundation stone of the Maison de la Chimie, planned as a world center for chemistry that would house a comprehensive chemical library, a conference center, offices for IUPAC and other bodies, and a documentation service that would maintain an inventory of publications from across the world and in all languages.[43] The Maison de la Chimie's global aspirations gave it the character of a monument that would honor a chemist of international standing while serving the interests of chemists throughout the world. The project's parallel, unwritten goal, however, was to advance France's standing in chemistry at a time when Germany was reemerging as a force, and hence a major rival, in the international chemical community.

Franco-German rivalry lent special poignancy to the similarity between the French project and the international institute of chemistry that Wilhelm Ostwald had hoped to establish in Brussels shortly before the war.[44] In discussions of the Maison de la Chimie, there was no mention of Ostwald, and the main promoters of the project— Jean Gérard, secretary-general of IUPAC and the French Société de chimie industrielle, and the Alsatian industrialist Paul Kestner—never entertained any other location for the institution than the French capital. Fund-raising too became charged with considerations of diplomacy and national interest. It was no coincidence that some of the largest contributions (in a total of almost twenty-five million francs, equivalent to about £1 million, or just under US$5 million, at contemporary rates) came from Latin America. Venezuela and Colombia, in particular, both anxious to affirm their place in the postwar world order, emerged as generous donors. In Europe, too, political motives loomed large. Here, the young nations of Czechoslovakia and Poland led the way with large donations that gave prominence on the stage of world science while cementing long-standing cultural ties with France. In contrast, Germany, though represented at the centenary and the source of three of the congratulatory addresses (among a total of 239) that were presented in the Sorbonne, gave virtually nothing. Britain, for its part, avoided the humiliation of what would have been a derisory contribution thanks only to a personal contribution of one million francs by the Francophile industrialist Sir Robert Mond.

Shot through with divergent national interests and in a financial climate that deteriorated menacingly in the early 1930s, the Maison de la Chimie did not get off to an easy start. Eventual installation in an elegant eighteenth-century building in the heart of Paris in 1934 certainly breathed new life into the project. But even then the institution's state-of-the-art library and information services had difficulty in winning recognition as the world resource that Gérard and Kestner always hoped they would become. Locally funded and more easily usable routes to the chemical literature, of the kind that the leading national societies provided through their own bibliographies and lending services, met chemists' needs better than a single depository. Also, and most damagingly for the project, in both Britain and the United States, perceptions of the Maison de la Chimie as an essentially French initiative, though never a source of outright opposition, fostered persistent coolness, even suspicion.[45]

The Pasteur and Berthelot centenaries, both of them large-scale commemorations of heroic figures and initiatives bordering on the realm of international diplomacy, were supreme examples of a favored French way of promoting the nation's scientific heritage. In other countries, celebratory practices tended to focus more heavily on the burgeoning genre of the museum of science and (in most cases) technology. The genre had deep roots, most notably, as it happens, in France in the collection of the Conservatoire des arts et métiers in Paris, dating from the 1790s and largely made up of objects confiscated after the Revolution. But in their modern twentieth-century form, they were largely a German creation, with collecting policies that embraced contemporary, often industrial, exhibits, as well as choice artifacts from the past. The most celebrated and influential of the new museums, the Deutsches Museum in Munich, was the work of Oskar von Miller, a distinguished Bavarian engineer and friend of Thomas Edison, known above all for his installation of a 109-mile high-voltage transmission line at the international electrical exhibition in Frankfurt in 1891. The line was a German triumph that marked a clear advance on the more modest low-voltage transmission demonstrated by the French engineer Marcel Deprez at the Munich electrical exhibition in 1882. It was one of the great "master works" that the museum set out to display.[46]

Miller's plan for the Deutsches Museum and the collections he assembled there made much of this and other achievements in Germany's rise as an industrial power. To that extent, the museum was an expression

8. Emperor Wilhelm II laying the foundation stone of the Deutsches Museum's permanent home on Coal Island in the River Isar, Munich, in 1906. Painting by Georg Waltenberger, 1916, now in the Deutsches Museum. The elaborate ceremony followed the decision to establish a museum of "masterpieces of science and technology" made during the annual meeting of the Verein Deutscher Ingenieure in 1903. Behind the emperor, holding papers, is the museum's leading promoter, Oskar von Miller, with other prominent engineers and industrialists, including Carl von Linde on his right. Photo Deutsches Museum (CD 66387), Munich.

of national pride. As such, it engaged the enthusiastic support of Crown Prince Ludwig of Bavaria from its foundation in 1903, and the blessing of the Kaiser, Wilhelm II, who laid the museum's foundation stone, amid the trappings of a great imperial occasion, three years later (figure 8). Despite the patriotic trappings, however, it was essential to Miller's vision of the museum's purpose that the scope of its collections should extend beyond a narrowly cast presentation of German achievements alone. National pride, for Miller, was empty bombast unless a display acknowledged the contribution of other nations as well.

Miller was true to his principle. But, as he knew, in museums with national status, the balance between parochialism and universal coverage was always difficult to strike. One distorting factor was that most modern artifacts were locally sourced, with disused teaching and demonstration apparatus and objects inherited from universal exhibitions often serving as key exhibits. As a result, the collections not only of the Deutsches Museum but also of other science museums that were inspired by Miller's example inevitably had something of a national profile. Displays at the National Technical Museum (Národni Technické Muzeum) in Prague and the Technical Museum (Technisches Museum für Industrie und Gewerbe) in Vienna, both founded in 1908, and the Science Museum in London, separated institutionally from the Victoria and Albert Museum in 1909, were all marked by their provenance.[47] What distinguished these prewar museums, however, was their capacity to fulfill political

9. Emperor of Austria Franz Josef laying the foundation stone of the Technisches Museum, Vienna, on June 20, 1909, shortly after celebrating his Golden Jubilee as emperor. Although work on the building began immediately, the museum did not open to the public until 1918. Courtesy of the Archive, Technisches Museum Wien.

expectations as symbols of an engagement with the modern world without being overly nationalistic in either tone or content (figure 9).

The contrast with displays that became increasingly common through the 1920s and 1930s was striking. Museum directors and their governmental paymasters were now far more likely to see a national museum as having a patriotic function that went beyond the purely informative and educational. It was a mark of the heightened perception of museums as showcases of national achievement that in 1926 two Swedish newspapers reported in some consternation that the country's contributions in science and technology were underrepresented in the Deutsches Museum's collections. As one of the newspapers had it, a survey of the museum revealed only a single object of Swedish origin: a pair of shoes from Lapland.[48] While such an observation was not decisive in setting the course for Stockholm's new museum of technology, the Tekniska Museet, it certainly provided supportive rhetoric for a venture that made much of its Swedish heroes. From its foundation in 1924, in fact, the Tekniska Museet was a product of the postwar "national turn" in that it made little attempt to develop an international dimension, preferring the mission of what was essentially a museum of Swedish industry and little else.[49]

In Britain, an even more explicit exercise in national image-building bore its first fruit in the British Empire Exhibition, held at Wembley in 1924 and then reopened in 1925. Here, the tone was international to the extent that the emphasis was on both the benefits that Britain had bestowed on its imperial possessions and the goods and services (not least in the First World War) that the dominions and colonies had made available to Britain.[50] The fact remained, however, that British achievements were at the heart of the project. In the Exhibition of Pure Science, housed close to but separately from the nearby Palaces of Industry and Engineering, Britain's contributions to modern physics, in particular those of J. J. Thomson, were given special prominence. It is telling that in the section on X-rays the work of their German discoverer, Wilhelm Röntgen, was reduced to the rank of "an almost chance observation" and quickly passed over in favor of Barkla's.[51]

National concerns were evident again in London in 1928, when a quintessentially British figure was honored at the Science Museum. With space made available by the inauguration of new galleries in the large East Block, the museum implemented an ambitious plan for the transfer to London of James Watt's workshop from Heathfield Hall, the house in Birmingham where it had lain virtually untouched since Watt's death in 1819.[52] It was no small consideration that the move satisfied the commercial interests of industrial sponsors on whose endorsement and material support the museum increasingly depended in coping with the economic difficulties of the 1920s. But British pride too was at work.

10. The Tempio Voltiano, Como, Italy, inaugurated in 1928 as a monument to Alessandro Volta. The temple was built to house Volta's few remaining instruments, with reproductions of many that had been lost in a fire during an exhibition in Como in 1899. Construction of the temple, designed by Federico Frigerio and financed by the industrialist Francesco Somaini, began in 1927, the year of the International Congress of Physicists in Como, Pavia, and Rome that marked the centenary of Volta's death. Copyright Archivio Nodo Como. Courtesy of the Fondazione Alessandro Volta, Como

Watt was the most enduring of all the nation's technological heroes, and the untidy, intensely workaday character of the workshop reinforced conventional perceptions of the practical man as the mainspring of Britain's industrial "take-off" in the eighteenth century.

Such exhibits, straddling the boundaries between culture and national interest, were part and parcel of postwar museological practices virtually everywhere. Italy was typical in its appropriation of the culture of science as the vehicle for a patriotically inspired assertion of modernity that drew on a glorious history of Latin intellect and culture. Following Mussolini's march on Rome in 1922 and the Fascist ascendancy that it precipitated, the country's scientific past was integrated into a patrimony that had hitherto been mainly expressed through artistic achievements, in particular of the Renaissance. The new emphasis on science was exemplified in the Tempio Voltiano in Como, completed in 1928 (one year late) to mark the centenary of Alessandro Volta's death (figure 10). In the manner of a Palladian shrine, the "temple" housed what remained of Volta's instruments, most of which had been destroyed in 1899 in a fire at that year's exhibition to celebrate the first experiments with the electric pile. It was in the same patriotic spirit that the magnificent instruments inherited from the Medici family, sanctified by their association with Galileo, were set at center stage in the Museo di Storia della Scienza, opened in the presence of Mussolini in Florence in 1930.[53]

In at least one respect, celebratory initiatives of this kind were encouraging, a sign that since the war science and technology had risen markedly in the priorities of governments intent on declaring their commitment to progressive values and human welfare. While this aspect was often cloaked in the rhetoric of the universality of knowledge and its benefits for all humanity, the universalist veil was generally thin and easily shed. The flamboyant procession through the streets of Munich on the occasion of the opening of the new premises of the Deutsches Museum in 1925, for example, was mounted as a show of confidence bred of Germany's accelerating reintegration in the community of nations (see figure 11). And similar assertiveness was evident in the museum's enlarged hall of scientific fame, the Ehrensaal, which found space only for the

11. Procession of floats in Munich, May 5, 1925, two days before the inauguration of the Deutsches Museum's new buildings on Oskar von Miller's seventieth birthday. Each float was devoted to a theme in the museum's displays. Here a float, topped with an aeroplane and drawn by seven Lanz "Bulldog" tractors, represents mechanical engineering. The procession and the warm reception it received in Munich conveyed the new confidence that marked postwar Germany's reintegration into the international community of nations. Photo Deutsches Museum (BN 31598), Munich.

portraits and busts of German scientists and engineers, a policy that remained intact until 1994, when Antoine-Laurent Lavoisier became the first non-German to be admitted (see figure 12).[54]

The totalitarian tide

The national turn in museums during the 1920s exemplifies the postwar challenge to perceptions of science as a body of truths independent of nation and local culture. But what occurred appears as little more than a mild adjustment by comparison with the thrust of German cultural policy under National Socialism. When Germany's new Chancellor

12. The Ehrensaal, Deutsches Museum, Munich, c. 1935. A hall of honor marking the achievements of leading scientists and engineers was part of Oskar von Miller's original plan for the museum. From its temporary initial premises in the old Bavarian National Museum, the Ehrensaal was enlarged and reinstalled in the museum's new buildings in 1925. Photo Deutsches Museum (BN 02487), Munich.

Adolf Hitler looked at the Deutsches Museum, still under Oskar von Miller's guiding hand, he found its tone and focus far too international; it did not convey strongly enough the genius of the German nation. By any normal standards, neither Miller nor the museum could be accused of a deficit of patriotic sentiment. But the standards that were imposed, once the passing of the Civil Service Restoration Act of April 1933 had lent the fig leaf of legitimacy to the dismissal of Jewish museum officials and academic professors, were of a new order. In a transformed climate, Miller quickly resigned from the Deutsches Museum's management committee, leaving a new administration to make such protective alliances as it could with the Nazi authorities.[55]

Hitler's preference was for the construction of a new museum of German technology, which he conceived as facing the Deutsches Museum on the opposite bank of the River Isar.[56] As it happened, under the pressure of other priorities the scheme never came to fruition, and it was left for the Deutsches Museum itself to adjust as best it could to the turn of political events. Acceptance of the spade that Hitler had used in inaugurating an Autobahn in Frankfurt am Main conveyed all too clearly the sacrifice of Miller's cultural norms in favor of political expediency.[57] But two special exhibitions in the museum's library signalled even more markedly the institution's capitulation to Nazi ideology. The first, inaugurated in November 1936, under the title "Der Bolschevismus. Grosse antibolschewistische Schau," was devoted to exposing the nature of Bolshevism and the Bolshevist threat.[58] The second, opened a year later and closely related to the first, was the notorious "Eternal Jew" ("Der ewige Jude") exhibition.[59] The two exhibitions were intended to alert a mass public to the threats posed by what they presented as related internationalist movements (the Bolshevist Jew Trotsky being a favored target on both counts). As the "Eternal Jew" exhibition set out to show, resistance to the age-old threat of Jewish infiltration in public life would only be possible if Germans of Aryan descent recognized the characteristics that distinguished the many Jews who, to right-wing nationalist concern, had risen to prominent positions in the arts and academic life during the Weimar Republic. Those characteristics included the physical attributes that were displayed, with the surface trappings of scientific objectivity, in the exhibition's first room, devoted to the "biological foundations of Judaism."[60]

In a display conceived with such anti-Semitic venom, science was a sad casualty. But the exhibition achieved its aims. Like the anti-Bolshevist exhibition, it was popular and received the endorsements of the National Socialist press and visits or shows of interest by party leaders, including Joseph Goebbels, Hermann Goering, and the architect of some of the most extreme anti-Jewish measures, Julius Streicher. Even if the exhibitions did not attract the numbers—some two million—that flocked to the better-known anti-Semitic exhibition of "Degenerate Art" ("Entartete Kunst"), held elsewhere in Munich between July and November 1937, they had the profile of cultural sensations.[61] In its first three weeks, attendances for the anti-Bolshevist exhibition reached 160,000, while "Eternal Jew" attracted an average of five thousand visitors a day over three months, nearly half a million in all, before going on to further success in Vienna and finally Berlin.

The context in which attempts to identify and promote what was truly German in science and culture is too well known to require detailed comment here. These were the years of the publication of a notorious preface to the four-volume manual of physics, *Deutsche Physik*, by Nobel Prize winner Philipp Lenard in 1936–1937 (see figure 13). The gratuitous insertion, in the preface, of a sharp contrast between Aryan and Jewish physics turned an otherwise innocuous and perfectly competent textbook of experimental physics into a work to be read as a manifesto for a "German Physics" that had no place for Einsteinian relativity and other supposedly Jewish contributions.[62] In a similar and far more menacing way, these were also the years of the emigration of large numbers of Jewish scientists expelled from their academic posts in accordance with pseudoscientific criteria of racial

Deutſche Phyſik

in vier Bänden

Von

Philipp Lenard

in Heidelberg

Allen, die in wohlgegründeter
Naturerkenntnis ihre geistige
Ruhe suchen, zur Freude ge-
schrieben.

Erſter Band:
Einleitung und Mechanik

Mit 135 Abbildungen

J. F. Lehmanns Verlag / München 1936

13. The title page of volume one of Philipp Lenard's Deutsche Physik in vier Bänden, 4 vols. (Munich: J. F. Lehmanns Verlag, 1936–1937). The work covered the whole range of experimental physics, with a minimum of mathematics. Although the fourth volume included a substantial section on modern physics, including cathode rays and radioactivity, it made no mention of relativity or quantum mechanics, which Lenard saw as products of Jewish physics. Deutsche Physik is now best known for the foreword to the first volume, dated August 1935, in which Lenard elaborated his insistence that only "Aryan physics" could be deemed truly German. Courtesy of the Science Museum/Science & Society Picture Library.

purity promoted in the Nuremberg laws from 1935, with their tragically familiar consequences.

By the time Nazi anti-Semitism exerted its baleful influence on any notion of the cosmopolitanism of science, there had already been disturbing premonitions elsewhere. One, in Italy, was the treatment of the eminent Jewish physicist Vito Volterra.[63] Volterra was anything but a political firebrand, and his patriotic credentials as a leader in his country's war effort during the First World War had been beyond question. In the spring of 1915, he had spoken forcefully in favor of Italy's passage from neutrality to an engagement on the Allied side. After joining the army corps of engineers in the same year (at age fifty-five), he had also been instrumental in persuading the Italian government to make systematic provision for weapons-related research in what soon became the Ufficio Invenzioni e Ricerche, the precursor of Italy's research-funding body, the Consiglio Nazionale delle Ricerche (CNR), established in 1923, with Volterra as its first president.[64] However, Volterra's concern for Italy's welfare had always gone hand in hand with a deeply rooted sense of citizenship in an undivided world of learning: he had been a prominent participant in the triennial international congresses of mathematicians from their foundation in Zurich in 1897 and, as a zealous traveller and correspondent, had maintained connections with disciplinary colleagues across Europe, especially in France and Britain. For someone so committed to transcending the boundaries of race and nation, the drift to dictatorial power under Mussolini raised acute tensions.

In 1931 the disparity between Volterra's internationalist commitment and the Fascist exploitation of science as an instrument of national aggrandisement contributed to his refusal to take the oath of allegiance to the king and the Fascist regime that was now required of university professors. It is a measure of the assimilation of Italian academic life in the apparatus of state that, out of a total of 1,250 professors, Volterra was one of only twelve who took such a stand. He duly lost his chair at the University of Rome, and three years later, following new legislation that obliged members of academies under state supervision to swear an oath of allegiance similar to that of 1931, he was expelled from the prestigious Accademia Nazionale dei Lincei. This came after more than forty-five years as a member of the Accademia, including four years as president in the 1920s.[65] The intended humiliation, though painful to him, did nothing to break his resolve or to diminish his standing in the international community. He retained his high offices in the IRC (from 1931, ICSU), the International Mathematical Union, and the International Committee of the Bureau international des poids et mesures, of which he remained president until his death in 1940.

In Spain too, scientists had to make decisions in response to intensifying authoritarian national sentiment. As a community, their sympathies lay generally with the republic and, from 1936, with the republican cause in the civil war. The leading Spanish physicist Blas Cabrera Felipe, a professor of electricity and magnetism at the University of Madrid who had spent a decisive period working on electromagnets with Pierre Weiss in the Zurich Polytechnikum (ETH) early in his career, was one who saw the Franquist dictatorship as incompatible with his vision of the openness of science.[66] His natural milieu was the one he frequented as a participant in the sixth Solvay Conference on Physics in 1930 and a prominent member of this and other international bodies, including the Bureau international des poids et mesures, which he served (eventually to the Franco regime's displeasure) as secretary from 1933 until 1941. As the leading Spanish physicist of his generation, his departure from Spain during the civil war and the life of exile that he led, first in Paris and then in Mexico, until his death in 1945, contributed substantially to the more general impoverishment of science under Franco.

Outstanding among those who shared Cabrera's determination to keep Spanish science open to the wider world was the neuroscientist and Nobel Prize winner (in 1906) Santiago Ramón y Cajal. Like Cabrera, Ramón y Cajal found a natural home in the Junta para Ampliación de Estudios e Investigaciones Científicas (JAE), a research organization, founded in 1907, whose strongly international focus embodied the liberal, modernizing spirit of Spain's Silver Age (Edad de Plata) in the first third of the twentieth century.[67] The JAE pursued its aims through scholarships for study abroad and the direction of funds to support research and ensure a strong Spanish presence in international congresses and collaborations. Depending as it did on public financing, however, it was always vulnerable to the chill winds that tended to constrain its activities under conservative administrations committed to more nationally focused agendas.

In the shifting political and economic climate of interwar Spain, the existence of the JAE would have been even more precarious had it not been for its association, from 1910, with the newly founded Residencia de Estudiantes in Madrid.[68] The Residencia was far more than the student residence that its name suggests. It housed a number of important laboratories, including Blas Cabrera's, in physics, and Pio de Rio-Hortega's, in histopathology. And visits by such figures as Marie Curie, Einstein, and Louis de Broglie helped to provide internationally minded members of Spain's scientific community with an access to the wider world of science that would have been difficult through other channels (figure 14). Although distinguished visiting lecturers left their mark across the range of the sciences, as well as in economics, history, and literature, it was in physics that contacts made in the Residencia had their greatest effect.[69]

It was the very openness of the Residencia, however, that was to make it an object of particular suspicion among supporters of the Franco regime, who saw it as a dangerously leftist interloper in the country's provision for learned culture. Constraints on the Residencia's activities from the beginning of the civil war in 1936 and the decision to close the JAE and transfer responsibility for research to the more easily controlled and intellectually less adventurous Consejo superior de investigaciones científicas (CSIC) when Franco consolidated his power three years later were all too predictable. They left Spanish intellectual life gravely impoverished and subject to what opponents of the dictatorship's domestically focused conception of science and scholarship could only interpret as a willful drift to isolation and a rejection of progressive values.

14. Blas Cabrera and Marie Curie in the Residencia de Estudiantes, Madrid. The photograph was taken during a visit to the Residencia in which Curie delivered a noted lecture on radioactivity on April 23, 1931. Cabrera directed the physics research laboratory in the Residencia and performed most of his research, on magnetism, there. Courtesy of the Residencia de Estudiantes, Madrid.

Through the 1930s, the appropriation of science and technology as tools of economic power and national propaganda gathered pace in most countries, democratic as well as totalitarian. The process was not without its material benefits, including in some instances enhanced allocations for research: the Department of Scientific and Industrial Research in Britain, the Caisse nationale des sciences (later the Caisse nationale de la recherche scientifique and finally, in 1939, the Centre national de la recherche scientifique) in France, and the CNR in Italy were typical of the governmental organizations that channelled much-needed funds into scientific and technological research between the wars. Prominent individual scientists too achieved the recognition that went with a greater appreciation of their value both for their material contributions and as symbols of national achievement. Guglielmo Marconi, who had pursued much of his research without institutional or governmental support, was one such beneficiary. The alliance he struck with the Italian Fascist Party, which he joined in 1923, and with Mussolini, who appointed him president of the Fascist-inspired Reale Accademia d'Italia in 1930, helped raise him to the status of a national hero worthy of the pomp of the state funeral that he received on his death in 1937.[70]

Even for someone as favored by the regime as Marconi, the path to such celebrity was not entirely smooth. Toward the end of his life, in particular, he found that proximity to government had its uncomfortable side. As someone with many close links abroad, he watched uneasily as Mussolini strengthened his alliance with Nazi Germany, embarked on colonial conquest in Abyssinia, and elaborated the aggressively anti-Semitic policies that culminated in the Italian racial laws of 1938, impeding many scientific careers, such as Volterra's, and leaving others with emigration as the only realistic option.[71] Enrico Fermi, who acted to protect his Jewish wife by leaving for the United States immediately on receiving his Nobel Prize for physics in Stockholm in 1938, was one who chose the path of exile, and it seems that Marconi himself may have contemplated a similar course.[72]

Totalitarianism, of course, could be of the Left as well as of the Right, and by now consequences of governmental intervention even darker than those in Italy were apparent in the Soviet Union. After the 1917 Revolution and on into the mid-1920s, Russian scientists had had no reason to see the Communist Party as a particular threat to their freedom; their problems in this period were mainly those of inadequate funding and social strife. But the ascendancy of Stalin in the late 1920s and the accompanying drives to industrialization and collectivization were the prelude to measures ranging from dismissals from key academic posts to imprisonment and execution.[73] As many scientists were to find to their cost, Stalinist authoritarianism went hand in hand with a new emphasis on national priorities and an incipient suspicion of Western science that was to have damaging consequences after the Second World War, notably in the promotion of Lysenkoism and the dismissal of mainstream genetics as a pseudoscience. Beginning in the mid-1930s and on irregularly into the early 1960s, the "cranks" (David Joravsky's term) in the discipline were the winners and science in its internationally accepted sense the loser.[74]

The rise of science in the hierarchy of national priorities everywhere made its mark on the universal exhibitions of the 1930s, never more flagrantly than at the International Exposition of 1937 in Paris. Preparations for the exhibition had begun in 1934, so that Léon Blum's left-wing Popular Front government inherited an event already well-planned when it came to power in May 1936. The Popular Front welcomed the opportunity of proclaiming its commitment to science-led modernity, expressed in the exhibition's official title, Exposition internationale des arts et techniques dans la vie moderne. To this end, it attached special importance to the unambiguously progressivist message conveyed in the Palais de la Découverte. Devoted to contemporary science, technology, and medicine, the Palais de la Découverte bore the thumbprint of the leading scientists on the political left who had conceived it: Jean Perrin, Paul Langevin, and Frédéric Joliot.[75] While it expressed their ambitions for the advancement of research in France, it also conveyed their strong commitment to internationalism, manifested in their collaboration with the leftist Confédération des travailleurs intellectuels and the French wing of the International Committee on Intellectual Co-operation. As Charlotte Bigg and Andrée Bergeron have argued, the two goals sat quite easily together.[76]

The Palais de la Découverte was generously provided for in a favored location in the western half of the Grand Palais, originally built for the Exposition Universelle of 1900. Its live demonstrations and imaginative Art Deco installation attracted 2,225,000 visitors in six months, representing an impressive share of the exhibition's total attendance of just over thirty-one million.[77] While the official report's view that the Palais was the "création capitale" of the exhibition owed something to the prevailing scientistic hyperbole of the day, it was not wide of the mark.[78] Despite the internationalist aspirations of Perrin, Langevin, and Joliot, the state of contemporary international politics gave the 1937 Exposition a confrontational edge that went beyond the

15. The German and Soviet pavilions at the Exposition internationale des arts et techniques dans la vie moderne in Paris, viewed from the terrace of the Trocadéro, 1937. The position of the pavilions (German on the left, Soviet on the right) set the stage for a blatant confrontation between the rival totalitarian powers. Copyright LAPI/Roger-Viollet.

usual competitiveness inherent in such events. This was most visibly manifested in the architecture of the various national pavilions, with most countries exploiting innovative design in fashioning the public face they presented to the world.

Two installations that caught every eye were the facing pavilions of the great totalitarian powers, Germany and the Soviet Union. The pavilions were designed with clear ideological intent, and both incorporated scientific and technical exhibits (among others) to communicate the optimistic modernism of their respective regimes.[79] In the German pavilion (on the left in figure 15), displays of achievement in engineering (strikingly illustrated by a huge Mercedes-Benz Zeppelin motor, an advanced sixteen-cylinder Mercedes racing car, and a show of fine optical instruments by Zeiss) were tempered with evocations, mainly through paintings, of healthy rustic simplicity. The intention was to show how the new Germany had succeeded in integrating technological sophistication with timeless German values, as opposed to those of international capitalism—a barely veiled repudiation of the unpatriotic capitalism of Jewish bankers and entrepreneurs. Across the exhibition's central avenue (ironically, the Avenue of Peace) stood the Soviet pavilion, topped with seventy-five-foot-high bronze statues of a peasant girl and a young industrial worker. There the emphasis was on the marshalling of technology for the common good, an ideal encapsulated in the choice of the mass-produced and stylistically unadventurous ZIS saloon car that looked down on the main exhibition hall. It was a presentation, like the building, with import quite as blatant as that of its Nazi counterpart. On either side of the Avenue of Peace, technology was given a distinctive national face.

Karen Fiss has argued convincingly that, internationalist rhetoric and ideals notwithstanding, Blum's government had master-minded, or at least condoned, what the visiting public viewed as raw, ideologically motivated confrontation. It even seems that a behind-the-scenes deal between the French organizers of the exhibition and the designer of the German pavilion, Albert Speer, had allowed Speer to adjust his plans and gain a six-meter advantage in height after being permitted, quite improperly, to see the design for the Soviet pavilion.[80] For the Popular Front, with its declared commitment to the anti-Fascist cause, an accommodation to Nazi wishes that verged on appeasement bore the seeds of political embarrassment, with the world as witness. In the event, whatever contact Speer may have had with the French authorities remained undisclosed, and the embarrassment did not materialize.

The potential of exhibitions as a genre that allowed national interests to be pursued behind the cloak of universalist virtue was not lost on the Fascist regime in Italy. Beginning in 1937, Mussolini laid the foundations for an exhibition in Rome that would be universal in name but a propaganda opportunity for the Fascist state in reality.[81] The Esposizione Universale di Roma was presented as an "Olimpiade delle civiltà," a setting for the fruits of the noblest aspirations of all humanity. In the words of the official presentation of the project, it would offer "a representation of the creative activities of the human spirit, wherever and however they manifest themselves amid the dynamism of that everlasting, fertile process of emulation on which there rests our assurance of universal progress, without boundaries or limits to its scope."[82] Planned to open on April 21, 1942, to coincide with the birth date of the city in 753 BC and mark the twentieth year of the Fascist era, E42 (the exhibition's abbreviated title) was intended to set new standards of opulence, as is evident to this day from some of the buildings that were erected before war curtailed the plan.

Like all universal exhibitions, E42 was about far more than science. But science had an important place. Monumental buildings around the exhibition's central square formed a "Città della Scienza" in which the development of scientific thought and the applications of science would be set among humanity's noblest achievements. The official rhetoric of the exhibition followed convention in making much of science as a pursuit unfettered by political boundaries: scientific and technical advances were to be presented as facets of a "Scienza Universale" that belonged to no one nation.[83] Nevertheless, the plans left privileged spaces for Italian heroes. In a projected Museum of the Sciences, Leonardo and Galileo were to have pride of place as the more distant representatives of a still vibrant national tradition in science, supremely encapsulated in modern times in the work of Marconi.[84]

An insistence on the undimmed quality of Italian scientific genius was essential to the profile of Fascist modernism that the exhibition was meant to convey. It found expression in the commission for the sculptor Arturo Dazzi to design a marble obelisk in honor of Marconi, to stand in the center of the piazza Imperiale, at the very heart of the exhibition. Although Dazzi began work, war interrupted the project, and the forty-five-meter obelisk that still stands in the renamed piazza G. Marconi was only completed in 1959, twenty years after the original commission. Since the exhibition did not take place, we cannot know how this and other facets of the Fascist-themed extravagance of E42 would have been received in countries unsympathetic to its blatantly political underpinnings. But even at the planning stage the gulf that separated E42's public face of harmony and world peace from the national interests that drove the project was evident. With the new racial laws aggravating the harassment of Volterra and other Jewish intellectuals, and Fermi beginning his new life in the United States, the planners' declarations of the universality of science had an especially sad irony.

The evidence that universalist ideals and national interests coexisted uneasily in preparations for the international exhibitions in Paris and Rome is unsurprising. More striking is that in a world edging once again to the threshold of war, the ideals found influential new voices. From the mid-1930s, H. G. Wells's conception of a "World Brain" that would act as a "mental clearing house" for humanity's collective mind revived hopes reminiscent of a more optimistic bygone.[85] How far Wells was influenced by the pre–First World War visionaries is uncertain. But there were clear echoes of the ideas of Otlet and Ostwald in his proposal for a "Permanent World Encyclopaedia" that lay at the heart of the World Brain idea.[86] As a limitless depository of the evolving stock of human culture, this "undogmatic Bible to a world culture," accessible to all and subject to constant revision, would help to sustain the broadly progressive pattern of human history that Wells had presented two decades before in his widely read *Outline of History.*[87] Internationalists of the generation of Otlet and Ostwald would have found much to approve in the project, the more so as Wells explicitly presented the supreme goal of the "Encyclopaedia" as the harmony between peoples that he believed would flow from the spread of understanding.

The fact remains, however, that as national interests asserted themselves with the particular force they acquired in the 1930s, the survival of internationalist ideals was no easy matter. Science and technology, in particular, had to struggle if they were to retain something of their old image as benign, transnational ventures wherever they existed. Such statements as, "In all ages science has been an international venture" and "The machine age owes its progress to all the workers of the world," included in the display of the French national committee of the ICIC at the Paris Exposition of 1937, expressed sentiments that would have gone virtually unquestioned in internationalist circles before 1914.[88] The difference now was that while the genuineness of the ICIC's commitment was never in doubt, such expressions of confidence in science and its applications no longer carried the easy conviction that they might have done a quarter of a century earlier. Lingering memories of the "chemists' war" had changed all that, as had the gathering reality of an arms race with military technology at its heart. Even if the characteristic ICIC dream of an "intellectual order" dedicated to humanity's "higher interests" still found a voice, by 1937 the voice had to struggle that much harder to be heard.[89] And within two years another onslaught on the world of mutual understanding was under way.

EPILOGUE

In September 1939, war once again interrupted the routines of intellectual life. Congresses had to be cancelled, and virtually all plans for academic visits and cultural exchanges abandoned. There were many casualties, institutional and personal. Among notable flagships of internationalism, the International Committee on Intellectual Co-operation and the network of national committees that sustained it succumbed quickly.[1] The German occupation of Paris in June 1940 precipitated the hasty transfer of the ICIC's staff and records to Bordeaux and the sealing of its Parisian premises by the German authorities. The Franco-German plan that took shape three months later allowed for a reformed ICIC to resume its activities from headquarters in Paris. But the new organization, now detached from the League of Nations (as the German authorities insisted), had no future other than as a vehicle for the spread of Nazi cultural propaganda, and even that aim was never fulfilled. As Jean-Jacques Renoliet has put it, the ICIC existed in a "deep coma" throughout the Occupation.[2]

It is consistent with an important theme of this book that, despite the impediments of war, the interests of the wider world of learning and an understanding of the bridge-building that the unimpeded movement of knowledge made possible were not totally lost from view. From 1942 to 1946, the British biochemist and sinologist Joseph Needham served as director of the Sino-British Science Co-operation Office, a post that involved him in extended wartime visits to unoccupied China, in particular to the office's headquarters in Chongqing, and opened his and other Western eyes to the potential for scientific collaboration with colleagues in a region of which even committed internationalists knew little.[3] More immediate in its effect was a three-day conference on "Science and world order," organized in September 1941 at the Royal

Institution in London under the auspices of the British Association for the Advancement of Science.[4] A gathering in which sessions were chaired by Edvard Beneš, president of Czechoslovakia, and the ambassadors of China, the Soviet Union, and the United States, was mounted with unambiguous strategic purpose, and virtually unqualified consensus was predictable. But, important though diplomacy and the bolstering of wartime alliances were, the presence of scientists from twenty-two nations and supportive messages from such eminent absent figures as Einstein, Harvard president James Conant, and the botanist V. L. Komarov, in his capacity as president of the Academy of Sciences of the USSR, lent the gathering ample intellectual respectability. When the president of BAAS, astronomer and former long-serving editor of *Nature* Sir Richard Gregory, presented a "Declaration of scientific principles" for approval at the concluding session of the conference, he was speaking to and for an audience convinced, like him, that the "true and great democracy" of science could only exist in a state of intellectual freedom.[5] Those German and Italian scientists who had allowed themselves to be seduced by the distorted racial principles of Nazi and Fascist intellectual life had thereby "renounced the international spirit of science" and (in the words of the British foreign minister Sir Anthony Eden) become "intellectually encircled."[6]

Despite the declaration's political edge, the debates at the Royal Institution showed how resilient the perception of science as a quintessentially universal culture remained, even in the darkest days. Once victory was achieved (at least in Europe), the perception resurfaced, and ideal moved toward reality with a promptness exemplified in the reemergence of the ICIC as early as February 1945. From the start, however, national interests resurfaced as well. In a manner familiar from the interwar years, the United States, with British support, campaigned for the creation of a completely new organization, one whose brief would be less elitist and extend beyond those who might be deemed "intellectuals."[7] An unspoken but unmistakable motive was the wish to break with an institution whose origins and administrative structure were thought to bind it too closely to France. In response, the French (led by the former Popular Front prime minister Léon Blum and Marie Curie's son-in-law Frédéric Joliot-Curie) fought to resurrect the ICIC along prewar lines. But at a conference attended by the representatives of forty-four states in London in November 1945, they finally admitted defeat and so paved the way for the establishment of UNESCO, the newly formed United Nations intergovernmental agency for educational and cultural matters.

The founding objectives of UNESCO, stated then and still today as "international peace and universal respect for human rights," were worthy of the most grandiloquent sentiments of an earlier age. So too was the high status that science occupied in the hierarchy of UNESCO's cultural priorities, despite at one stage coming close to being omitted from the organization's brief altogether. Crucial to the prominence of science was the contribution of two leading scientists with strong international commitments: the biologist Julian Huxley, UNESCO's first director-general, and Joseph Needham, fresh from his mission to China, who became the first head of its scientific section. In the event, neither man stayed long. Both were gone within two years, following conservative disapproval of their left-wing political views and, in Huxley's case, the prominent role he had played in British humanism.

Since then, other political and cultural priorities have often worked against UNESCO's internationalist objectives. In particular, the capricious way in which the United Kingdom and the United States have at various times withheld and then resumed their support has conveyed a lack of sustained commitment that has repeatedly exasperated the organization's supporters. Despite the pitfalls, however, UNESCO's core guiding principles remain in place. In some respects, in fact, advances in Internet and electronic communication have breathed new life into them, in particular where the free exchange of information (a leading UNESCO priority) is concerned.

A conspicuous beneficiary has been the long-frustrated dream of global access to the sum of human knowledge, with its roots stretching back long before the First World War. Here, private ventures, rather than UNESCO-sponsored initiatives, have tended to lead the way. The

Google Books Library Project, launched in 2004 from its precursor in the more modest Google Book Search, was one such venture. At the time, the project's initial plan of digitizing fifteen million volumes and making them available online (4.5 billion pages in all) appeared exhilaratingly bold. The objections, though, soon multiplied. Some focused on the poor quality of the images and legal problems to do with copyright. Other, more fundamental, objections concerned questions of selection and scope. Since even the colossal number of titles proposed for digitization represented barely a tenth of those published since the birth of printing, it was obvious that choices would have to be made. Who then would do the choosing? If the task was left to Google, a US-based company with strong commercial priorities, the choices risked being systematically skewed toward works in English. Then again, how would scholarly criteria fare in competition with pressures that might favor titles likely to maximize the hits on the Google website and so attract advertisers?

The attack on the project, led by Jean-Noël Jeanneney, a former president of the Bibliothèque nationale de France, provoked second thoughts on the part of several of the participating libraries that had agreed to make books available for scanning. For Jeanneney, outlining what he presented in 2007 as "a view from Europe," the Google initiative had aspired to an ideal of "universal knowledge" but delivered a chimera, a "myth."[8] Although the number of digitized items in the Google database went on to top thirty million (twice the original target), the criticisms struck home, and progress with the project in recent years has been more circumspect.[9] Encouragingly, however, the project's difficulties have stimulated other initiatives. An Internet portal administered from The Hague, the Europeana, has launched a multinational platform giving access to European resources, Anglophone and non-Anglophone, in twenty-seven countries, with a target of having thirty million items available by 2015.[10] And in 2013, the Digital Public Library of America (DPLA) finally got off the ground, not as a centralized digital enterprise but rather, in the manner of the Europeana, as a pathway leading via more than forty hubs to a vast heritage of printed ephemera, archives, and artifacts as well as books.[11]

So, might new techniques of information science and the World Wide Web have set us at last on our way to the goal of knowledge open to all that inspired the work of the Institut international de bibliographie or Wells's unrealized "World Encyclopaedia? " Arguably, they have. The fact that some important digital resources (such as the HathiTrust Digital Library and Early English Books Online) are still only available to subscribing institutions underlines how far we have yet to go. But the more distant prospect continues to beckon enticingly and evoke ever more ambitious responses. One of the most recent of these, commonly described as the Global Brain, follows the Europeana and DPLA in constructing a network of distributed but interconnected sources of knowledge rather than a centralized fount of information.[12] As its advocates insist, the Global Brain's dispersed focus opens the door to an essentially democratic intelligence in which the interactions among the totality of human beings, augmented in recent years by social media, have a constitutive role. That fluidity sets the Global Brain apart from Wells's plan for a World Brain, whose rigidity and implicit endorsement of the collectivist "new world order" in which Wells saw humanity's greatest hope of salvation from social tensions and warfare have always disquieted critics.[13] The Global Brain's self-organizing character distances it too from Otlet's ambitions for his Mundaneum as a central repository of the world's accumulated body of information. As a devotee of order, Otlet would probably have found the Global Brain disturbing. Yet he would have fully endorsed its overriding goals. And it is not hard to imagine the personal satisfaction he would have derived in 2012 from the choice of the Free University of Brussels, his own university, as the location for the world's only Global Brain Institute, devoted to research in information and communication technology. In the work of that distinguished center and its promise of "ever-stronger connectivity between peoples,"[14] the Otlet story might even be said to have come full circle.

NOTES

Introduction

1 Cf. the journal's full title: *Philosophical Transactions: Giving Some Accompt of the Present Undertakings, Studies, and Labours of the Ingenious in Many Considerable Parts of the World.*

2 Anne Rasmussen, "Jalons pour une histoire des congrès internationaux au XIXe siècle: régulation scientifique et propagande intellectuelle," *Relations internationales* 62 (1990): 115–133; and "L'Internationale scientifique 1890–1914," doctoral thesis, Ecole des Hautes Etudes en Sciences Sociales, Paris, 1995.

3 I have occasion to refer to Schroeder-Gudehus's work at various points in the book. Her *Les Scientifiques et la paix: La communauté scientifique internationale au cours des années vingt* (Montreal, 1978) remains a classic contribution.

4 In making this distinction, I follow Daniel Laqua, who discusses the work of Otlet and La Fontaine in the chapter on "Universalism" in Laqua, *The Age of Internationalism and Belgium 1880–1930: Peace, Progress and Prestige* (Manchester and New York, 2013), 181–210. I am grateful to Tara Windsor for a discussion of this and other key secondary sources in the course of the workshop "Science en guerre & guerre des savants: Politique et réseaux scientifiques internationaux durant la première guerre mondiale," Académie royale de Belgique, November 13–14, 2014.

5 Laqua, *Age of Internationalism*, 21. The words quoted are from Piérard's *Wallons et Flamands* (Brussels, 1929), 33. The notion of a "commerce des idées" appears in the preface to the first edition of Henri Pirenne's *Histoire de Belgique* (Brussels, 1900), vol. 1, viii–ix; also reproduced in later editions.

6 Louis Frank, *Les Belges et la paix* (Brussels, 1905), 178. For a comment, see Laqua, *Age of Internationalism*, 22.

7 Laqua, *Age of Internationalism*, 35.

8 Ibid., 1–16 ("Introduction").

9 Martin H. Geyer and Johannes Paulmann, eds., *The Mechanics of Internationalism: Culture, Society, and Politics from the 1840s to the First World War* (Oxford, 2001); Madeleine Herren, *Hintertüren zur Macht: Internationalismus und modernisierungsorientierte Außenpolitik in Belgien, der Schweiz und den USA 1865–1914* (Munich, 2000); Madeleine Herren and Sacha Zala, *Netzwerk Aussenpolitik: Internationale Kongresse und Organisationen als Instrumente der schweizerischen Aussenpolitik 1914–1950* (Zurich, 2002); and Glenda Sluga, *Internationalism in the Age of Nationalism* (Philadelphia, 2013).

10 Geyer and Paulmann, "Introduction: The Mechanics of Internationalism," in Geyer and Paulmann, *Mechanics of Internationalism*, 1–25 (esp. 2).

11 For references to the work of Lindqvist and Somsen, see chapter 1, notes 35 and 77, and chapter 3, note 48.

12 Hermann Kellermann, ed., *Der Krieg der Geister: Eine Auslese deutscher und ausländischer Stimmen zum Weltkriege 1914* (Weimar, 1915).

13   On the striking examples of William Ramsay in Britain and Pierre Duhem, see chapter 2, 48–49.

14   In addition to the sources cited in chapter 3, Madeleine Herren and Sacha Zala offer a particularly clear overview of the appropriation of internationalist ideology by totalitarian regimes engaged in the organization of exhibitions and other cultural initiatives during the Nazi period; see their *Netzwerk Aussenpolitik*, 151–190. The authors' Swiss perspective makes for a distinctive analysis in a now abundant literature.

15   Robert W. Rydell, *World of Fairs: The Century-of-Progress Expositions* (Chicago and London, 1993). Studies of the individual exhibitions include Lisa D. Schrenk, *Building a Century of Progress: The Architecture of Chicago's 1933–34 World's Fair* (Minneapolis and London, 2007); and, on the New York World's Fair, Robert H. Kargon, Karen Fiss, Morris Low, and Arthur P. Molella, *World's Fairs on the Eve of War: Science, Technology and Modernity, 1937–1942* (Pittsburgh, 2015), 57–82.

16   On these Europe-wide academies, founded in 1988 and 2003, respectively, see their websites at http://www.ae-info.org/ and http://www.eurasc.org/.

17   Rasmussen, "L'Internationale scientifique, 1890–1914."

Chapter 1

1   What I say about the persistence of the aspiration to the possession of universal knowledge is informed by a very helpful comment by Anita Guerrini. As Professor Guerrini has observed, the absurdity of Edward Casaubon's consuming aspiration to write a history of all mythology, in George Eliot's *Middlemarch* (1871–1872), reflects contemporary perceptions of the impossibility of such an all-encompassing venture.

2   A comment quoted in Isaac Todhunter, ed., *William Whewell, D.D., Master of Trinity College, Cambridge: An Account of his Writings with Selections from his Literary and Scientific Correspondence*, 2 vols. (London, 1876), vol. 1, 410.

3   Augustin Boutaric, *Marcellin Berthelot 1827–1907* (Paris, 1927), 188–189. A similar point emerges strongly from obituaries published at the time of Berthelot's death. See, for example, the notice in the *Times*, March 19, 1907.

4   Although Poincaré's core interests remained rooted in mathematics, philosophical and ethical concerns informed many of his writings, of which *La Science et l'hypothèse* (Paris, 1902) in particular had a broad nonmathematical readership; see Jeremy Gray, *Henri Poincaré: A Scientific Biography* (Princeton, NJ, and Oxford, 2013), esp. chapter 1 ("The Essayist"), 27–152. For E. T. Bell, Poincaré was the "last universalist" in mathematics; see the chapter bearing that title in Bell, *Men of Mathematics* (London, 1937), 583–612.

5   Ann Blair, *Too Much to Know: Managing Scholarly Information in the Modern Age* (New Haven, CT, and London, 2010).

6   Ibid., 78.

7   Ben Jonson, *Timber or Discoveries Made upon Men and Matter*, ed. Felix E. Schelling (Boston, 1892), 7.

8   For statistics concerning the production of books and periodicals, I draw on Paul Otlet's essay on what he variously called "bibliologie" or "documentologie" in Otlet, *Traité de documentation : Le livre sur le livre : Théorie et pratique* (Brussels, 1934), 9–42. The words expressed Otlet's aim of promoting bibliography as part of a true science of documentation.

9   *Nomenclator autorum omnium, quorum libri vel manuscripti, vel typis expressi exstant in Bibliotheca Academiae Lugduno-Batavae* (Leiden, 1595).

10   The first Bodleian catalogue (1605) is reproduced, with a helpful introduction, in *The First Printed Catalogue of the Bodleian Library 1605. A Facsimile. Catalogus librorum Bibliothecae publicae quam vir ornatissimus Thomas Bodleius Eques Auratus in Academia Oxoniensi nuper instituit* (Oxford, 1986).

11   See the striking pie chart representing the distribution of the languages of incunabula at schools-wikipedia.org/images/2489/248927.png.htm. The figure of 70 percent for Latin in this first half century of printing far surpassed those for German (10.8 percent), Italian (8 percent), and French (5.7 percent). The proportion did not exceed 2 percent for any other language, the figure for English standing at 0.8 percent.

12   Charles Fantazzi, "Harriot's Latin," in Robert Fox, ed., *Thomas Harriot and His World: Mathematics, Exploration, and Natural Philosophy in Early Modern England* (Farnham and Burlington, VT, 2012), 231–236.

13   Felicity Henderson, "Faithful Interpreters? Translation Theory and Practice at the Early Royal Society," *Notes and Records of the Royal Society* 67 (2013): 101–122 (esp.106–117).

14   Adrian Johns, *The Nature of the Book: Print and Knowledge in the Making* (Chicago and London, 1998), 514–521.

15   Henderson, "Faithful Interpreters?": 112–113. As Henderson shows, the pressure to translate work in the more recondite languages, such as Dutch and Danish, was even greater.

16   In his account of the gradual abandonment of Latin for scientific purposes during the eighteenth century, however, Michael Gordin stresses the complexity of a process that saw the Swedish chemist Torbern Bergman reverting to Latin in communicating his ideas in the 1770s and early 1780s; see Gordin, *Scientific Babel: The Language of Science from the Fall of Latin to the Rise of English* (London, 2015), 41–48. Here and elsewhere, my references are to the Profile Books edition of Gordin's book, which has the same pagination as the University of Chicago Press edition, also published in 2015, though with a different subtitle.

17   Between 1799 and 1802, 34 percent of the papers read in the academy's meetings were delivered in Latin, compared with 44 percent in French, 15 percent in Russian, and 6 percent in German. See "Histoire de l'Académie impériale des sciences. Années MDCCXCIX–MDCCCII," in *Nova acta academiae scientiarum imperialis Petropolitanae* 15 (1806): 42–60.

18   Otlet, *Traité de documentation*, 38–39.

19 Derek de Solla Price, *Science since Babylon*, enlarged edition (New Haven and London, 1975), 163–170.

20 Alex Csiszar, "Seriality and the Search for Order: Scientific Print and Its Problems during the Late Nineteenth Century," *History of Science* 48 (2010): 399–434.

21 Price, *Science since Babylon*, 176n.

22 Graeme J. N. Gooday, *The Morals of Measurement: Accuracy, Irony, and Trust in Late Victorian Electrical Practice* (Cambridge, 2004), 59–60.

23 Richard Taylor, ed., *Scientific Memoirs, Selected from the Transactions of Foreign Academies of Science and Learned Societies, and from Foreign Journals*, 5 vols. (London, 1837–1852). On Taylor's venture and his internationalist credentials, see Maeve Olohan, "Gate-Keeping and Localizing in Scientific Translation Publishing: The Case of Richard Taylor and Scientific Memoirs," *British Journal for the History of Science* 47 (2014): 433–450; and W. H. Brock and A. J. Meadows, *The Lamp of Learning: Two Centuries of Publishing at Taylor & Francis*, 2nd ed. (London, 1998), 102–106. Taylor also edited, printed, and published the long-established *Philosophical Magazine*.

24 Roy M. MacLeod, "Evolutionism, Internationalism and Commercial Enterprise in Science: The International Scientific Series 1871–1910," in A. J. Meadows, ed., *The Development of Science Publishing in Europe* (Amsterdam, 1980), 63–93; Leslie Howsam, "An Experiment with Science for the Nineteenth-Century Book Trade," *British Journal for the History of Science* 33 (2000): 187–207; and Valérie Tesnière, *Le Quadrige: Un siècle d'édition universitaire 1860–1968* (Paris, 2001), 48–52 and 82–83.

25 John Culbert Faries, *The Rise of Internationalism* (New York, 1915), 41–96. The high proportion of congresses in science, medicine, and technology (almost half of the total number listed) is conveyed clearly in Robert Doré, *Essai d'une bibliographie des congrès internationaux* (Paris, 1923), 2–56. For a modern study, see Anne Rasmussen, "Jalons pour une histoire des congrès internationaux au XIXe siècle: régulation scientifique et propagande intellectuelle," *Relations internationales* 62 (1990): 115–133.

26 Brigitte Schroeder-Gudehus, "Division of Labour and the Common Good: The International Association of Academies, 1899–1914," in Carl Gustaf Bernhard, Elisabeth Crawford, and Per Sörbom, eds., *Science, Technology, and Society in the Time of Alfred Nobel* (Nobel Symposium 52; Oxford, 1982), 3–20 (esp. 3).

27 Brigitte Schroeder and Anne Rasmussen, *Les Fastes du progrès : Le guide des expositions universelles 1851–1992* (Paris, 1992), 138–139, for a helpful list.

28 *Institut de France. Académie des sciences : Congrès astrophotographique international tenu à l'Observatoire de Paris pour le levé de la carte du ciel (avril 1887)* (Paris, 1887). Important contributions on the history of the project are assembled in part I ("Historical Research") of Suzanne Débarbat et al., eds., *Mapping the Sky* (Dordrecht, Boston, and Paris, 1988), 9–148. An informative brief study is Théo Weimer, *Brève histoire de la Carte du ciel* (Paris, 1987). For a participant's account, see Herbert Hall Turner, *The Great Star Map: Being a*

Brief General Account of the International Project Known as the Astrographic Chart (London, 1912).

29 Faltering did not imply abandonment of the project, however. Work continued sporadically after the First World War, and despite the organizational and technical difficulties, some of the surviving photographic plates, as well as data from the Astrographic Catalogue, have been used in recent research, notably in conjunction with results from the European Space Agency's Hipparcos satellite.

30 Terry Quinn, *From Artefacts to Atoms: The BIPM and the Search for Ultimate Measurement Standards* (Oxford, 2012), passim but esp. 74–110.

31 Peter Alter, "The Royal Society and the International Association of Academies 1897–1919," *Notes and Records of the Royal Society* 34 (1979–1980): 241–264; and Brigitte Schroeder-Gudehus, "Division of Labour and the Common Good" and the same author's "Die Akademie auf internationalem Parkett: die Programmatik der internationalen Zusammenarbeit wissenschaftlicher Akademien und ihr Scheitern im Ersten Weltkrieg," in Jürgen Kocka et al., *Die Königlich Preussische Akademie der Wissenschaften, Die Königlich Preussische Akademie der Wissenschaften zu Berlin im Kaiserreich* (Berlin, 1999), 175–195.

32 Robert Marc Friedman, *The Politics of Excellence: Behind the Nobel Prize in Science* (New York, 2011), 59–65.

33 Adolf Erik Nordenskiöld, ed., *Vega-expeditionens vetenskapliga iakttagelser bearbetade af deltagare i resan och andra forskare*, 5 vols. (Stockholm, 1882–1887). On polar exploration as a quintessentially Swedish science, see Tore Frängsmyr, "Swedish Polar Science," in Frängsmyr, *Science in Sweden: The Royal Swedish Academy of Sciences 1739–1989* (Canton, MA, 1989).

34 On Arrhenius, I draw on Elisabeth Crawford, *Arrhenius: From Ionic Theory to the Greenhouse Effect* (Canton, MA, 1996).

35 Geert J. Somsen, "A History of Universalism: Conceptions of the Internationality of Science from the Enlightenment to the Cold War," *Minerva* 46 (2008): 361–379 (esp. 365–367).

36 The special status of Belgium as a "site of internationalism" is explored in detail in Daniel Laqua, *The Age of Internationalism and Belgium, 1880–1930* (Manchester and New York, 2013); see esp. 1–16 for a succinct summary of Laqua's argument.

37 In an extensive and growing literature on Otlet and La Fontaine, W. Boyd Rayward's *The Universe of Information: The Work of Paul Otlet for Documentation and International Organisation* (Moscow, 1975) remains indispensable. Valuable more recent studies of Otlet include Françoise Levie, *L'Homme qui voulait classer le monde : Paul Otlet et la Mundaneum* (Brussels, 2006); and Alex Wright, *Cataloging the World: Paul Otlet and the Birth of the Information Age* (New York, 2014). La Fontaine's work is well treated in a handsomely produced volume of essays: *Henri La Fontaine, Prix Nobel de la Paix: Un Belge épris de justice* (Mons and Brussels, 2012).

38 *Conférence bibliographique internationale : Bruxelles 1895. Documents* (Brussels, 1895). I follow Rayward in using the title of the IIB to describe

what were in reality two related institutions. The other institution, the Office international de bibliographie, functioned essentially as the administrative channel for the subsidy that the work of the IIB received from the Belgian government.

39 Rayward, *Universe of Information*, 179–195; Levie, *L'Homme qui voulait classer le monde*, 121–123; and Wright, *Cataloging the World*, 116–123.

40 YouTube (https://www.youtube.com/watch?v=p4QJMSeo_cI) has brief footage showing Carnegie being shown the museum by La Fontaine and a particularly attentive Otlet.

41 Levie, *L'Homme qui voulait classer le monde*, 149; and Wright, *Cataloging the World*, 118.

42 Rayward, *Universe of Information*, 192.

43 Wright, *Cataloging the World*. Cf. the similar sense of the title of Françoise Levie's book on Otlet, *L'Homme qui voulait classer le monde* (cited above, note 37).

44 The objectives are well conveyed in Paul Otlet and Ernest Vandeveld, *La Réforme des bibliographies nationales et leur utilisation pour la Bibliographie universelle : Rapport présenté au Ve Congrès international des éditeurs* (Milan, 1906).

45 Thomas Hapke, "Wilhelm Ostwald, the 'Brücke' (Bridge), and Connections to Other Bibliographic Activities at the Beginning of the Twentieth Century," in Mary Ellen Bowden, Trudi Bellardo Hahn, and Robert V. Williams, eds., *Proceedings of the 1998 Conference on the History and Heritage of Science Information Systems* (Medford, NJ, 1999), 139–147.

46 A still valuable account of Ostwald's life and work is the article on him by Erwin N. Hiebert and Hans-Günther Körber in Charles Coulston Gillispie, ed., *Dictionary of Scientific Biography*, 16 vols. (New York, 1970–1980), vol. 15, 455–469.

47 As described in Ostwald, *Die Weltformate: I. Für Drucksachen* (Munich, 1912); and Karl Wilhelm Bührer, *Raumnot und Weltformat* (Munich, 1912).

48 Wilhelm Ostwald, *Das Gehirn der Welt* (Munich, 1912).

49 Elke Behrends, *Technisch-wissenschaftliche Dokumentation in Deutschland von 1900 bis 1945 unter besonderer Berücksichtigung des Verhältnisses von Bibliothek und Dokumentation* (Wiesbaden, 1995), 19–28.

50 *Catalogue of Scientific Papers (1800–1863): Compiled and Published by the Royal Society of London* [and varied titles], 19 vols. (London and Cambridge, 1872–1925). On the origins of the project, see "Preface," vol. 1, iii–vi.

51 For an account of the early history of the ICSL, see the preface to *International Catalogue of Scientific Literature. First Annual Issue: A. Mathematics* (London, 1902), v–ix. The same preface appeared in the corresponding volumes for the other subject areas.

52 See E. J. Crane's essay, "Chemical Abstracts," in Charles Albert Browne and Mary Elvira Weeks, *A History of the American Chemical Society: Seventy-Five Eventful Years* (Washington, DC, 1952), 336–367.

53 What eventually became *Physics Abstracts* began in 1898 as *Science Abstracts: Physics and Electrical Engineering*, a joint venture of the Institution of Electrical Engineers, the Physical Society of London, and the American Physical Society.

54 *Chemical Abstracts Service (CAS) June 14, 2007: National Historic Chemical Landmark*, a commemorative booklet published on the occasion of the centenary of the service and its designation as a National Historic Chemical Landmark in 2007. Also available online at acs.org/content/acs/en/education/whatischemistry/landmarks/cas.html.

55 Although Sarton was sixteen years younger than Otlet, and with interests more rooted in the sciences, accounts of his student days in Brussels in the early years of the twentieth century point to his immersion in a heritage of late-nineteenth-century progressive thought similar to that which Otlet encountered as a law student there in the 1880s. See Arnold Thackray and Robert K. Merton, "On Discipline-Building: The Paradoxes of George Sarton," *Isis* 63 (1972): 472–495 (esp. 477–482); and the same authors' article on Sarton in Gillispie, *Dictionary of Scientific Biography*, vol. 12, 107–114.

56 Roger Fennell, *History of IUPAC 1919–1987* (Oxford, 1994), 3–10; Danielle Fauque, "French Chemists and the International Reorganisation of Chemistry after World War I," *Ambix* 58 (2011): 116–135; and Brigitte Van Tiggelen and Danielle Fauque, "The Formation of the International Association of Chemical Societies," *Chemistry International* 34, no. 1 (January-February 2012): 8–11. For a contemporary account, see the unsigned "Historique de l'Union internationale de la chimie pure et appliquée," in *Union internationale de la chimie pure et appliquée : Comptes rendus de la première conférence internationale de la chimie. Rome 22–24 juin 1920* (Paris, n.d.), 3–8.

57 "Association internationale des sociétés chimiques. Extrait du procès-verbal de la seconde session du Conseil," *Archives des sciences physiques et naturelles*, 4th ser. 33 (1912): 524–530; and, more briefly, "International Association of Chemical Societies," *Nature* 89 (May 9, 1912): 245–246.

58 See Fennell, *History of IUPAC*, 10, for a list of the national delegates (each with three votes) and the nonvoting members from three affiliated societies: the Deutsche Bunsengesellschaft, the Faraday Society, and the French Société de chimie physique. Solvay's internationalism had much in common with Otlet's, including a liberal, scientistic streak that also seems to have owed something to his exposure to the contemporary tide of Comtean positivism; see Kenneth Bertrams, Nicolas Coupain, and Ernst Homberg, *Solvay: History of a Multinantional Family Firm* (Cambridge, 2013), 100–101 and 140–141.

59 Wilhelm Ostwald, *Denkschrift über die Gründung eines Internationalen Instituts für Chemie* (Leipzig, 1912).

60 "Institut chimique international," a news item in *La Vie internationale* 1, fascicule 1 (1912): 116.

61 I follow Michael Gordin in using the terms "auxiliary" and "ethnic" rather than the commonly used "artificial" and "natural." See Gordin, *Scientific Babel*, 109–114.

62 On the rise and decline of Russian as a scientific language in the second half of the nineteenth century, with the reputations of the chemists Alexander Butlerov and Dmitri Mendeleev as a main theme, see Gordin, *Scientific Babel*, 51–103.

63 Despite the obvious difficulties, determined but unsuccessful attempts to establish French as the international language of intellectual communication were made in the nineteenth and early twentieth centuries, notably at two conferences to promote the cause, in Liège in 1905 and Arlon three years later. See F. S. L. Lyons, *Internationalism in Europe 1815–1914* (Leiden, 1963), 208.

64 The movement is well described in Gordin, *Scientific Babel*, 105–158.

65 Ibid., esp. 118–148. For a detailed account of the world Esperanto movement, with brief discussions of rival auxiliary languages before the First World War, see also Peter G. Forster, *The Esperanto Movement* (The Hague, Paris, and New York, 1982), 1–168.

66 Carl Bourlet, "Le huitième Congrès universel d'Esperanto. Cracovie 1912.08.11/18," *La Vie internationale* 1, fascicule 5 (1912): 563–573.

67 Forster, *The Esperanto Movement*, 45–49; and Gordin, *Scientific Babel*, 114–118 and 143–144.

68 On the conception, content, and eventual demise of Ido, see Gordin, *Scientific Babel*, 131–158.

69 Louis Couturat, *A Plea for an International Language* (London, 1903), 4.

70 Ibid., 29.

71 Volker Welter, *Biopolis: Patrick Geddes and the City of Life* (Cambridge, MA, 2002), esp. 54–112; Levie, *L'Homme qui voulait classer le monde*, 147–148; and Wright, *Cataloging the World*, 108–123.

72 On Andersen's life, with special reference to his friendship with James, see Millicent Bell's introduction to the English edition and the translation of Rosella Mamoli Zorzi's introduction to the original Italian edition of Henry James, *Beloved Boy: Letters to Hendrik C. Andersen 1899–1915*, ed. Rosella Mamoli Zorzi (Charlottesville, VA, and London, 2004), xxxvii–lv. Andersen housed his collection in the Villa Helene, designed and named to commemorate his mother, Helene Monsine Monsen, from 1924. He bequeathed the house to the Italian government on his death in 1940, and the property became a museum in 1999.

73 Dario Matteoni stresses this aspect in his essay on "L'ideologia del pacifismo e la città," in Giuliano Gresleri and Dario Matteoni, *La Città mondiale: Andersen, Hébrard, Otlet, Le Corbusier* (Venice, 1982), 11–105.

74 As Volker Welter has observed, Andersen's plan for a single "world centre" contrasted with Geddes's vision of a "world league" of cities distributed across the globe. See Welter, *Biopolis*, 76.

75 Hendrik Christian Andersen, *Creation of a World Centre of Communication by Hendrik Christian Andersen. Ernest M. Hébrard Architect* (Paris, 1913).

76 Ibid., 32–44.

77 Svante Lindqvist, "An Olympic Stadium of Technology: Deutsches Museum and Sweden's Tekniska Museet," in Brigitte Schroeder-Gudehus, ed.,

Industrial Society and Its Museums 1890–1990 (Chur, 1993), 37–54 (esp. 39–40). A lavishly produced report on every detail of the Games conveys Sweden's determination that the Fifth Olympiad should be, in the words of the president of the Swedish Olympic Committee, Prince Gustav Adolf, "the greatest international trial of strength in the athletic field that our times can show." For this comment, see Erik Bergvall, *The Fifth Olympiad: The Official Report of the Olympic Games of Stockholm 1912. Issued by the Swedish Olympic Committee*, trans. Edward Adams-Ray (Stockholm, 1913), 310.

78   Robert H. Kargon and Arthur P. Molella, *Invented Edens: Techno-Cities of the Twentieth Century* (Cambridge, MA, and London, 2008); also the account of Andersen's city in Matteoni, "L'ideologia del pacifismo e la città."

79   Andersen, *Creation of a World Centre*, 45–46.

80   Ibid., 45–76, esp. 69–76.

81   The quotations are from two letters of frank criticism from James to Andersen, April 14, 1912, and September 4, 1913, in James, *Beloved Boy*, 101–103 and 110–112.

82   Andersen sent the copy "printed for the University of Oxford, Oxford, England" with an explanatory letter, to the chancellor of the university, Lord Curzon, who passed the book on to the Bodleian Library (Bod. 17364 b.4). The copy in the British Library (K.T.C.III.b.8) records that the volume had been "printed for the British Museum, London, England." The French edition in the Bibliothèque nationale de France, one of seventy-five printed on Japan paper, is marked as "imprimé pour M. Raymond Poincaré Président de la République française" [BnF RES G-Z115 (1)].

83   Flammarion to Andersen, quoted from an undated letter in *"World Conscience": An International Society for the Creation of a World-Centre to House International Interests and Unite Peoples and Nations for the Attainment of Peace and Progress upon Broader Humanitarian Lines* (Rome, 1913), 38.

84   Ibid., 27–48, for the other endorsements.

85   La Fontaine and Otlet to Andersen, February 28, 1912, letter reproduced in *"World Conscience,"* 30–31.

86   The figures for attendances at the world congresses are taken from Otlet's typewritten "Historique de l'Union des associations internationales," UAI OP – dossier 97, Mundaneum Archives, Mons.

87   *La Vie internationale : Revue mensuelle des idées, des faits et des organismes internationaux* 1, fascicule 5 (1912): 619–623. *La Vie internationale* was published in Brussels by the OCAI. Five volumes appeared between 1912 and 1914, and one further issue was published in November 1921. A similarly titled predecessor of the *Annuaire* had appeared since 1905 under the auspices of the Institut international de la paix, the creation (in 1903) of Albert I of Monaco. It became an official publication of the UAI and adopted its definitive title in 1910, with the aid of the newly founded Carnegie Endowment for International Peace.

88   A point well made in David S. Patterson, "Andrew Carnegie's quest for world peace," *Proceedings of the American Philosophical Society* 114 (1970): 371–383.

Chapter 2

1 *Traité de paix générale basé sur une charte mondiale déclarant les droits de l'humanité et organisant la confédération des états* (Brussels, 1914). Although the author's name did not appear on the title page, a signed preface ("Présentation"), dated October 10, 1914, leaves no doubt that the work was by Otlet.

2 Henri La Fontaine, *The Great Solution. Magnissima Charta: Essay on Evolutionary and Constructive Pacifism* (Boston, 1916).

3 The origin of Haber's comment is hard to identify. But in a valedictory statement posted on the notice board of the Kaiser Wilhelm Institute for Physical Chemistry and Electrochemistry on the occasion of his resignation from the KWI in 1933, he expressed his pride in an institution that, in the twenty-two years of his directorship, had been "dedicated to serving humanity in times of peace, and the fatherland in times of war"; see Daniel Charles, *Between Genius and Genocide: The Tragedy of Fritz Haber, Father of Chemical Warfare* (London, 2005), 232–233. I am grateful to Joseph Gal for drawing this source to my attention.

4 A succinct overview of science's role in the First World War is Jon Agar, *Science in the Twentieth Century and Beyond* (Cambridge and Malden, MA, 2012), 89–117. Among more detailed studies, see the essays in Jeffrey Johnson and Roy M. MacLeod, eds., *Frontline and Factory: Comparative Perspectives on the Chemical Industry at War* (Dordrecht, 2006). The essays in *Le Sabre et l'éprouvette: L'invention d'une science de guerre 1914/1939*, in the series *14–18 Noesis*, no. 6 (Paris, 2003), 39–172, include contributions on subjects other than chemistry.

5 L. F. Haber, *The Poisonous Cloud: Chemical Warfare in the First World War* (Oxford, 1986), 239–258; Olivier Lepick, *La Grande Guerre chimique: 1914–1918* (Paris, 1998), 314–321; Kim Coleman, *A History of Chemical Warfare* (Basingstoke and New York, 2005), 11–38.

6 For these early exchanges, see Hermann Kellermann, ed., *Der Krieg der Geister: Eine Auslese deutscher und ausländischer Stimmen zur Weltkrieg 1914* (Weimar, 1915).

7 Published as "Aufruf 'An die Kulturwelt,'" in the *Berliner Tageblatt*, October 4, 1914, and very quickly in ten languages. On the composition and impact of the manifesto, see Jürgen and Wolfgang Von Ungern-Sternberg, *Der Aufruf "An die Kulturwelt!"* (Stuttgart, 1996), with the original German text (including preliminary drafts) and French and English translations on pp. 156–164. A contemporary French translation, with a helpful introduction by A. Morel-Fatio, is in *Les Versions allemande et française du manifeste des intellectuels allemands dit des quatre-vingt-treize*, 2nd ed. (Paris, 1915). Documents illustrating the reciprocal abuse leading up to the "Aufruf" are in Kellermann, *Der Krieg der Geister*, 27–113.

8 William Ramsay, "Germany's Aims and Ambitions," *Nature* 94, no. 2345 (October 8, 1914): 137–139 (esp. 138).

9 Pierre Duhem, *La Science allemande* (Paris, 1915), passim but esp. (and

succinctly) 103–143; and "Science allemande et vertus allemandes," in Gabriel Petit and Maurice Leudet, eds., *Les Allemands et la science* (Paris, 1916), 137–152 (esp. 138–145). For a helpful discussion of Duhem's treatment of style in science as a reflection of national mentality, see Harry W. Paul, *The Sorcerer's Apprentice: The French Scientist's Image of German Science 1840–1919* (Gainesville, FL, 1972), 54–76.

10 Stefan L. Wolff, "Physiker im 'Krieg der Geister': die 'Aufforderung' von Wilhelm Wien," *Acta historica Leopoldina* 48 (2007): 41–62.

11 "Aufruf an die Europäer," in Nicolai, *Die Biologie des Krieges: Betrachtungen eines Naturforschers den Deutschen zur Besinnung*, 2 vols. (Zurich, 1917), vol. 1, 12–16. The text appeared in English as "A Manifesto to Europeans," in Nicolai, *The Biology of War*, trans. Constance A. Grande and Julian Grande (London, 1919), 7–9; reproduced in David E. Rowe and Robert Schulmann, eds., *Einstein on Politics: His Private Thoughts and Public Stands on Nationalism, Zionism, War, Peace, and the Bomb* (Princeton, NJ, and Oxford, 2007), 64–66.

12 The influence of the prehistory of antagonism, dating back to the Franco-Prussian War of 1870, is an important theme of Harry Paul's *Sorcerer's Apprentice*, esp. 1–28.

13 "Report of Council, 1918," in *Year-Book of the Royal Society of London: 1919* (London, 1919), 171–180 (esp. 172). The proposal was forwarded to the first of the inter-allied conferences, held in London in October 1918 (on which, see text, 57–58). But at the conference no action was taken, the measures taken against the Central powers being considered sufficient.

14 Einstein praised the stand taken by Planck and Fischer in his letter to Romain Rolland, September 15, 1915; see Rowe and Schulmann, *Einstein on Politics*, 71–72.

15 Donna Ewald and Peter Clute, *San Francisco Invites the World: The Panama-Pacific International Exposition of 1915* (San Francisco, 1991).

16 *Exposition universelle et internationale de San Francisco : La Science française*, 2 vols. (Paris, 1915).

17 On the debates surrounding the awards during the First World War, see Robert Marc Friedman, *The Politics of Excellence: Behind the Nobel Prize in Science* (New York, 2001), 71–92; and "Has the Swedish Academy of Sciences … Seen Nothing, Heard Nothing, and Understood Nothing?" in Rebecka Lettevall, Geert Somsen, and Sven Widmalm, eds., *Neutrality in Twentieth-Century Europe: Intersections of Science, Culture, and Politics after the First World War* (New York and London, 2012), 90–114. In the sciences, the only Nobel Prize to be presented during the war was that of Robert Bárány; see note 26.

18 These words, from Nobel's will, were incorporated into the statutes of the Nobel Foundation. See "Code of Statutes of the Nobel Foundation. Given at the Palace in Stockholm, on the 29th day of June in the year 1900," in Elisabeth Crawford, *The Beginnings of the Nobel Institution: The Science Prizes, 1901–1915* (Cambridge and Paris, 1984), 221–229 (esp. 221–222).

19 "Special regulations, concerning the distribution etc. of prizes from the Nobel Foundation by the Royal Academy of Science in Stockholm. Given by

His Gracious Majesty, Oscar II, King of Sweden and Norway, at the Palace in Stockholm, on the 29th day of June 1900," ibid., 230–235 (esp. 230).

20 Crawford, *Beginnings of the Nobel Institution*, 109–149; and Friedman, *Politics of Excellence*, 40–53. On Arrhenius, an essential source is Elisabeth Crawford, *Arrhenius: From Ionic Theory to the Greenhouse Effect* (Canton, MA, 1996).

21 Crawford, *Beginnings of the Nobel Institution*, 126–128; Crawford, *Arrhenius*, 174–175 and 235–237; and Friedman, *Politics of Excellence*, 180–184.

22 See chapter 1, 23.

23 Elisabeth Crawford, *Nationalism and Internationalism in Science, 1880–1939: Four Studies of the Nobel Population* (Cambridge, 1992), 61–64.

24 My comments on the award to Haber and the reactions to it draw mainly on Friedman, *Politics of Excellence*, 93–115.

25 Ibid., 113–114.

26 The only other winner of a scientific prize during the war, the Austro-Hungarian otologist, Robert Bárány, who had been awarded the 1914 prize for physiology or medicine, had exceptionally received his prize at a Nobel ceremony in 1916.

27 See http://www.nobelprize.org/nobel_prizes/physics/laureates/1917/barkla-speech.html for Barkla's speech at the banquet.

28 Brian Wynne, "C. G. Barkla and the J Phenomenon: A Case Study in the Treatment of Deviance in Physics," *Social Studies of Science* 6 (1976): 307–347.

29 Frank Greenaway, *Science International: A History of the International Council of Scientific Unions* (Cambridge, 1996), 19–32. Concise accounts of the conferences and the decisions taken at them are in the *Year-Book of the Royal Society of London: 1919* (London, 1919), 181–184, and the corresponding volume for 1920 (London, 1920), 176–181. .

30 Determination of what constituted an appropriate "adhering body" was a complex matter. The normal preference was for a principal national academy or national research council. But the definition of "national membership" of what was a nongovernmental organization raised difficulties that persist in ICSU to today. On the resulting diversity of practices, see the useful historical sketch in *Reports of Proceedings of the General Assembly of the International Council of Scientific Unions*, vol. 5 (1951?): 113–117; also Sir Henry Lyons, *The Royal Society 1660–1940: A History of Its Administration under Its Charters* (Cambridge, 1944), 316–320; and Greenaway, *Science International*, 23–24 and 36–41.

31 The sixteen adhering bodies were from Australia, Belgium, Brazil, Canada, France, Greece, Italy, Japan, New Zealand, Poland, Portugal, Romania, Serbia, South Africa, the United Kingdom, and the United States. On Russia as a particularly conspicuous absentee, see this chapter, 73, and chapter 3, 76 and 129, note 4.

32 In the Union académique internationale and its associated bodies, the linguistic constraint was even more severe, French alone being allowed.

33 Roswitha Reinbothe, *Deutsch als internationale Wissenschaftssprache*

und der Boykott nach dem Ersten Weltkrieg (Frankfurt-am-Main, 2006), 35. Reinbothe draws on statistics in D. B. Baker, F. A. Tate, and R. J. Rowlett, Jr., "Changing Patterns in the International Communication of Chemical Research and Technology," *Journal of Chemical Documentation* 11 (1971): 90–98 (esp. 91).

34   John L. Heilbron, *The Dilemmas of an Upright Man: Max Planck and the Fortunes of German Science*, new ed. (Cambridge, MA, and London, 2000), esp. 69–81.

35   For evidence that Planck and Fischer had not seen the text of the manifesto before agreeing to sign, see the letter from Einstein to H. A. Lorentz, August 2, 1915, in Rowe and Schulmann, *Einstein on Politics*, 69–70.

36   At Planck's request, Lorentz circulated the letter to chosen recipients. I quote from the English translation that appeared in *The Observatory: A Monthly Review of Astronomy* 39 (June 1916): 284–285. For a comment on the letter, accompanying a slightly different translation, see Heilbron, *Dilemmas of an Upright Man*, 76–78.

37   See Einstein's letters of July 21 and August 2, 1915 to Lorentz, in Rowe and Schulmann, *Einstein on Politics*, 68–70.

38   See chapter 3, 85.

39   Carl Bourlet, "Le huitième congrès universel d'Esperanto," *La Vie internationale* 1, fascicule 5 (1912): 563–573 (esp. 565–566).

40   Terry Quinn, *From Artefacts to Atoms: The BIPM and the Search for Ultimate Measurement Standards* (Oxford, 2012), 194–198.

41   Ibid., 197. The point about illegality is well made in Brigitte Schroeder-Gudehus, "Probing the Master Narrative of Scientific Internationalism: Nationals and Neutrals in the 1920s," in Rebecka Lettevall, Geert Somsen, and Sven Widmalm, eds., *Neutrality in Twentieth-Century Europe: Intersections of Science, Culture, and Politics after the First World War* (New York and London, 2012), 19–42 (esp. 22).

42   Although the statutes of the URSI were not formally approved until the IRC's General Assembly in 1922, they came into effect immediately on the Union's foundation in 1919.

43   Wolfgang Torge, "The International Association of Geodesy 1862 to 1922: From a Regional Project to an International Organization," *Journal of Geodesy* 78 (2005): 558–568.

44   Felix Schmeidler, *Die Geschichte der Astronomischen Gesellschaft: Jubliäumsband 125 Jahre Astronomische Gesellschaft* (Hamburg, 1988), esp. 3–26.

45   Ibid., 42–50. On the Bonner Durchmusterung, see Alan H. Batten, "Argelander and the Bonner Durchmusterung," *Journal of the Royal Astronomical Society of Canada* 85 (1991): 43–50.

46   Evan Hepler-Smith, "'Just as the Structural Formula Does': Names, Diagrams, and the Structure of Organic Chemistry at the 1892 Geneva Nomenclature Congress," *Ambix* 62 (2015): 1–28. See also Hepler-Smith, "Nominally Rational: Systematic Nomenclature and the Structure of Organic Chemistry, 1889–1940" (Princeton University, PhD thesis, 2016). I am grateful to Dr. Hepler-Smith for showing me two draft chapters from the thesis, in

advance of submission. Valuable older sources include Maurice P. Crosland, *Historical Studies in the Language of Chemistry* (London, 1962), 347–354; Pieter Eduard Verkade, *A History of the Nomenclature of Organic Chemistry*, trans. S. G. Davies (Dordrecht, Boston, and Lancaster, 1985), 1–73; and Roger Fennell, *History of IUPAC 1919–1987* (Oxford, 1994), 1–10.

47 Verkade, *History of the Nomenclature of Organic Chemistry*, 75–87.

48 Joseph W. Dauben, "Mathematicians and World War I: The International Diplomacy of G. H. Hardy and Gösta Mittag-Leffler as Reflected in their Personal Correspondence," *Historia Mathematica* 7 (1980): 261–288; and Crawford, *Arrhenius*, 242–243.

49 Yoshiyuki Kikuchi, "World War I, International Participation and Reorganisation of the Japanese Chemical Community," *Ambix* 58 (2011): 136–149 (esp. 140–146); and James R. Bartholomew, *The Formation of Science in Japan: Building a Research Tradition* (New Haven, CT, and London, 1989), 254–263. I refer to Sakurai and Tanakadate by their family names and give those names first, in accordance with Japanese practice. Tanakadate adopted the given name Aikitu rather than Aikitsu, although the latter sometimes appears in the secondary literature.

50 Kenkichiro Koizumi, "The Emergence of Japan's First Physicists, 1868–1900," *Historical Studies in the Physical Sciences* 6 (1975): 3–108 (esp. 72–81).

51 Olivia Cushing Andersen and Hendrik Christian Andersen, *Creation of a World Centre of Communication by Olivia Cushing Andersen and Hendrik Christian Andersen. Legal Argument from the Positive Science of Government by Umano. The Economic Advantages by Prof. Jeremiah W. Jenks* (Rome, 1918).

52 Ibid., "Preface," i–iii.

53 Published in Italian as Umano [pseudonym of Gaetano Meale], *Positiva scienza di governo. Per parlare di politica senza vecchie ciurmerie e asinerie …* (Turin, 1922).

54 Andersen's letter to Curzon, dated February 1919, is tipped into the copy of the supplementary volume, now in the Bodleian Library (Bod. 24885 b.1).

55 The copy of the French edition of the supplement that Andersen sent to Poincaré (now in the Bibliothèque nationale de France, BnF RES G-Z-115[2]), bears a characteristic handwritten dedication: "S. E. le Président de la République Française. Témoignant ma reconnaissance et la plus haute estime. Hendrik Christian Andersen. Roma. 1919." A typewritten letter to Poincaré, dated February 1919, accompanies this copy. In it Andersen expresses the hope that his ideas might contribute to the establishment of a "Centre Mondial Administratif" for a future Society of Nations. I am grateful to Paul H. Thomas of the Hoover Institution Library and Archives for drawing my attention to copies of the 1913 and 1918 volumes in his care, each with a dedication to Hoover in Andersen's hand, dated 1921.

56 Françoise Levie, *L'Homme qui voulait classer le monde : Paul Otlet et le Mundaneum* (Brussels, 2006), 165–176, brings out well the effect of the loss of Otlet's son on his attempts to take up his activities after the war.

57 For a glowing contemporary description, with photographs, see *Centre*

international. Conceptions et programme de l'internationalisme. Organismes internationaux et Union des associations internationales. Établissements scientifiques installés au Palais Mondial (August 1921), unpaginated.

58 W. Boyd Rayward, *The Universe of Information: The Work of Paul Otlet for Documentation and International Organisation* (Moscow, 1975), 192 and 241.

59 Rayward, *Universe of Information*, 268; Levie, *L'Homme qui voulait classer le monde*, 204; and Alex Wright, *Cataloging the World: Paul Otlet and the Birth of the Information Age* (New York, 2014), 119 and 158.

60 Rayward, *Universe of Information*, 221–229; and Levie, *L'Homme qui voulait classer le monde*, 189–191. The richest source of documents concerning the university is the Mundaneum Archives, Mons; see especially the files PP PO 211-dossier 9 ("Université internationale 1920") and Publications UAI 06. Of numerous printed announcements of the university's courses, the substantial brochure *Sur la création d'une université internationale : Rapport présenté à l'Union des associations internationales* (Brussels, February 1920) is informative.

61 Among precedents and possible models for the international university, see in particular the somewhat similar proposal, incorporating study in a number of different countries, by the American geologist Angelo Heilprin. The proposal is outlined in Heilprin, "The Ignorance of Education and the Project of an International University," *The Forum* 29 (March 1900): 71–78.

62 *L'Université internationale. Première session. 5–20 septembre 1920 : Programme des cours et conférences* (Brussels: Palais Mondial [Cinquantenaire], 1920) ; and *L'Université internationale. Documents relatifs à sa constitution. Rapport—conférence—statut. Session inaugurale. Bruxelles. Palais mondial. Publication no. 1. 1920* (Brussels, 1920), esp. 5–7 and 109–124, on the procedural formalities, the numbers attending, and the courses given.

63 "La quinzaine internationale" and "L'Université internationale," *La Vie internationale*, fascicule 26, "No 1 post bellum" (November 1921): 137–195.

64 Rayward, *Universe of Information*, 239 and 254.

65 A printed announcement of the session planned for 1923 is in the file PP PO 224-dossier 1, Mundaneum Archives, Mons. There is no evidence that the session took place.

66 Rayward, *Universe of Information*, 261–271; Levie, *L'Homme qui voulait classer le monde*, 203–208; and Wright, *Cataloging the World*, 173–174. A letter of July 6, 1924, from Otlet to the president and members of the International Committee on International Co-operation, explaining the threat and asking for support, is in the Paul Otlet papers, UAI 229, Mundaneum Archives, Mons.

67 Wright, *Cataloging the World*, 167–174.

68 Initially "Mondaneum," but quickly changed to "Mundaneum."

69 *Biological Abstracts* was founded in 1926 from the union of *Abstracts of Bacteriology* and *Botanical Abstracts*, both of which had been established a decade or so earlier.

70 Rayward, *Universe of Information*, 274–303. On Otlet's worsening relations with Duyvis, see Wright, *Cataloging the World*, 202–203.

71 Rayward, *Universe of Information*, 255–259; Levie, *L'Homme qui voulait*

classer le monde, 194–203; and Wright, *Cataloging the World*, 177–179. For an early expression of Otlet's indignation at what he saw as the marginalization of the UAI by the ICIC, see *La Société des Nations: Rapport aux Associations sur les premiers actes de la Commission de Coopération Intellectuelle, par M. Paul Otlet, secrétaire général de l'Union. Janvier 1923*. These twenty-eight pages of sustained invective were composed to arouse resentment among members of the UAI.

72 For lists of the early members of the committee (twelve initially, rising to seventeen by 1930), see Jan Kolasa, *International Intellectual Cooperation (The League Experience and the Beginnings of UNESCO)* (Wrocław, 1962), 167–171; and Henri Galabert, *La Commission de Coopération intellectuelle de la Société des Nations* (Toulouse, 1931), 50–53 and 72–78.

73 See Jean-Jacques Renoliet, *L'UNESCO oubliée : La Société des Nations et la coopération intellectuelle (1919–1946)* (Paris, 1999) for a fine study of the ICIC and the wider context of interwar ventures in intellectual cooperation. Kolasa, *International Intellectual Cooperation*, which similarly presents the ICIC's place as a precursor of UNESCO, is another valuable source. Galabert, *La Commission de Coopération intellectuelle* is an illuminating contemporary account.

74 Martin Kohlrausch and Helmuth Trischler, *Building Europe on Expertise: Innovators, Organizers, Networkers* (Basingstoke, 2014), 97–102; and Frank Hartmann, "Visualizing Social Facts: Otto Neurath's ISOTYPE Project," in W. Boyd Rayward, ed., *European Modernism and the Information Society: Informing the Present, Understanding the Past* (Aldershot and Burlington, VT, 2008), 279–293. On Neurath's association with Otlet and his establishment of a Vienna Mundaneum in 1931, see Nader Vossoughian, *Otto Neurath: The Language of the Global Polis* (Rotterdam and London, 2008), 96–113; and Wouter Van Acker, "Internationalist Utopias of Visual Education: The Graphic and Scenographic Transformation of the Universal Encyclopaedia in the Work of Paul Otlet, Patrick Geddes, and Otto Neurath," *Perspectives on Science* 19 (2011): 32–80.

75 Rayward, *Universe of Information*, 345–363; and Wright, *Cataloging the World*, 203–204.

76 Paul Otlet, *Traité de documentation. Le livre sur le livre : Théorie et pratique* (Brussels, 1934). The dates of the beginning and completion of the book's production, in 1932 and 1934 respectively, appear on the reverse of the title page: "Commencé d'imprimer 1932.00. Achevé d'imprimer 1934.04."

77 Otlet's "Postface" was published unpaginated, with relevant correspondence, at the end of the *Traité de documentation*.

78 Levie, *L'Homme qui voulait classer le monde*, 311–318. The salvaging of what remained of the card catalogues and records of the IIB and UAI has had a delayed but happy sequel in their rehousing in Mons, in a handsome museum and research center, appropriately named the Mundaneum. The Mundaneum (website at www.mundaneum.org) has been open to scholars and the general public since 1998.

79 Quoted in Wright, *Cataloging the World*, 179.

80 Andersen expressed his indignation in an incensed letter to Otlet, June 4, 1929, protesting against "the corruption of my original and organic plans";

ibid., 188. For the plan, see Paul Otlet, Le Corbusier, and Pierre Jeanneret, *Mundaneum* (Union des associations internationales, publication no. 128; Brussels, 1928); also the accounts in Wright, *Cataloging the World*, 179–189; and Giuliano Gresleri's essay "Da Bruxelles a Ginevra: aforismi e avanguardia architettonica," in Giuliano Gresleri and Dario Matteoni, *La Città mondiale: Andersen, Hébrard, Otlet, Le Corbusier* (Venice, 1982), 107–255, esp. 125–131. The attack on the project by the avant-garde Czech artist Karel Teige is discussed in Kohlrausch and Trischler, *Building Europe on Expertise*, 79–94.

81 The copy sent to Henderson, inscribed with a handwritten dedication, is now in the London Library.

82 Hendrik C. Andersen, *World-Conscience; An International Society for the Creation of World Peace by the Establishment of a World Centre City of Communication Conceived by Hendrik Christian Andersen* (Rome, 1934), 10. On Andersen's interview with Mussolini, see Millicent Bell's "Introduction" to Henry James, *Beloved Boy: Letters to Hendrik C. Andersen 1899–1915*, ed. Rosella Mamoli Zorzi (Charlottesville, VA, and London, 2004), ix–xxxv (esp. xxxi).

83 Andersen, *World-Conscience* (1934), 10.

84 Bell, "Introduction," in James, *Beloved Boy*, xxxi; also the chronological timeline in the same volume, 149.

Chapter 3

1 Paul Weindling, "Public Health and Political Stabilisation: The Rockefeller Foundation in Central and Eastern Europe between the Two World Wars," *Minerva* 31 (1963): 253–267; and "The Rockefeller Foundation and German Biomedical Sciences, 1920–40: From Educational Philanthropy to International Science Policy," in Nicolaas A. Rupke, ed., *Science, Politics and the Public Good: Essays in Honour of Margaret Gowing* (Basingstoke and London, 1988), 118–140.

2 George E. Vincent, "President's Review" in *The Rockefeller Foundation: Annual Report. 1922* (New York, 1923), 7–66 (esp. 37).

3 For evidence of a particularly generous allocation of these awards in 1926, see Richard M. Pearce, "Division of Medical Education," *The Rockefeller Foundation: Annual Report 1926* (New York, 1927), 343–356 (esp. 346–350).

4 The Soviet Union never joined the IRC and only joined its successor, ICSU, in 1954, a decision precipitated by Soviet interest in the forthcoming International Geophysical Year of 1957–1958. On the importance of personal friendships, with special reference to mathematicians, see Christopher Hollings, *Scientific Communication across the Iron Curtain* (Cham, 2016), 1–20.

5 Paul Weindling, "German-Soviet Medical Co-Operation and the Institute for Racial Research, 1927–c1935," *German History* 10 (1992): 177–206 (esp. 177–189).

6 P.-V. Angus Leppan, "A Note on the History of the International Association of Geodesy," *Journal of Geodesy* 58 (1984): 224–229 (esp. 225–226);

Wolfgang Torge, "The International Association of Geodesy 1862 to 1922: From a Regional Project to an International Organization," *Journal of Geodesy* 78 (2005): 558–568 (esp. 566–567); and Adriaan Blaauw, *History of the IAU: The Birth and First Half-Century of the International Astronomical Union* (Dordrecht, 1994), 62–67.

7   On the foundation and early history of the IAU, see Blaauw, *History of the IAU*, 1–102.

8   Roger Fennell, *History of IUPAC 1919–1987* (Oxford, 1994), 26–29.

9   T. A. C. Schoevers, ed., *Report of the International Conference of Phytopathology and Economic Entomology: Holland 1923* (Wageningen, 1924 [?]). In all, twenty-five countries were represented, and papers were given in English, German, and French. As the editor of the Report pointedly observed, "a spirit of mutual esteem and good-understanding prevailed," and "not a single dissonant note was heard" (p. 5).

10   K. Jordan and W. Horn, eds., *Verhandlungen des III. Internationalen Entomologen-Kongresses. Zürich, 19–25. Juli 1925*, 2 vols (Weimar, 1926).

11   Ibid., vol. 1, 55–72, for a list of the national delegations and others attending.

12   "Zur Vorgeschichte," ibid., vol. 1, 11–12.

13   The phrase (in French) is in the address of welcome to the congress by the president of the Swiss Entomological Society, Arnold Pictet; ibid., vol. 1, 26.

14   Blaauw, *History of the IAU*, 76–80.

15   Roger Adams, "William Albert Noyes 1857–1941," *National Academy of Sciences: Biographical Memoirs* 27 (1952): 179–208 (esp. 188 and 192).

16   An apparently trenchant letter (now lost) in which Noyes expressed his concern to the president of IUPAC, William Pope, is referred to in a letter of March 18, 1924, to Pope from the IUPAC secretary general Jean Gérard; IUPAC Archives (CHF), Series II. Bureau Correspondence, 1924–1927. Box 4.

17   Quoted in the prefatory material to the proceedings of the Bologna Congress of 1928: *Atti del Congresso Internazionale dei Matematici: Bologna 3–10 settembre 1928 (VI)*, 6 vols. (Bologna, 1929–1932), vol. 1, 5.

18   See the two pages of published minutes of the 1924 General Assembly of the IMU, in J. C. Fields, ed., *Proceedings of the International Mathematical Congress Held in Toronto, August 11–16, 1924*, 2 vols. (Toronto, 1928), vol. 1, 65–66 (esp. 66).

19   Pincherle made his unsuccessful plea to Picard in a letter of June 8, 1928, three months before the Bologna congress; see *Atti del Congresso Internazionale dei Matematici: Bologna*, vol. 1, 8–9n.

20   The unsigned text in which Pincherle's letter is reproduced, titled "Preparazione" (ibid., 5–19), conveys the veiled acrimony of the exchanges that culminated in the issuing of invitations to mathematicians from the former Central powers. The irregular manner in which the Bologna congress had been planned, as what was effectively an independent Italian initiative, resulted in a parting of the ways between the IMU, seen as irredeemably hard-line in its adherence to the IRC's statutes, and an overwhelming majority in

the international mathematical community. The result was an uncoupling of the IMU from the running of future congresses and, following the failure of attempted compromises at the General Assembly during the Zurich congress of 1932, its liquidation until the formation of a new IMU after the Second World War; see Olli Lehto, *Mathematics without Borders; A History of the International Mathematical Union* (New York, 1998), 23–71; and Guillermo P. Curbera, *Mathematicians of the World, Unite! The International Congress of Mathematicians: A Human Endeavor* (Wellesley, MA, 2009), 75–90.

21 Frustration at the IRC's hard line engendered particular bitterness in 1925, following the defeat, at that year's IRC General Assembly, of a Norwegian proposal (supported by other Scandinavian countries, the United States, and Britain) for the lifting of the ban on the admission of the Central powers. See Sir Henry Lyons, *The Royal Society 1660–1940: A History of its Administration under its Charters* (Cambridge, 1944), 318.

22 The Federal Republic of Germany joined the IRC's successor, ICSU, with the Deutsche Forschungsgemeinschaft (DFG) as its representative body, in 1952. The German Democratic Republic, represented by the Deutsche Akademie der Wissenschaften zu Berlin, followed in 1961. The DFG is now the sole German representative.

23 Paul Forman, "Scientific Internationalism and the Weimar Physicists: The Ideology and Its Manipulation in Germany after World War I," *Isis* 64 (1973): 150–180. Forman contrasts the prevailing internationalism of Weimar's literary and artistic culture with the reluctance of a majority in the German physics community to engage in the formal process of reconciliation.

24 Picard made the point in his opening address at the General Assembly of the IRC in 1928. See *International Research Council. Fourth Assembly held at Brussels, July 13th, 1928: Reports of Proceedings edited by Sir Arthur Schuster, F.R.S., General Secretary* (London, 1928), 1–4.

25 Fennell, *History of IUPAC*, 29–32. The efforts of Cohen and Biilmann bore other delayed fruit in Austria's admission to IUPAC in 1932. Hungary, however, never joined, pleading that the country's financial circumstances prevented involvement in any of the unions.

26 Frank Greenaway, *Science International. A History of the International Council of Scientific Unions* (Cambridge, 1996), 36–38.

27 On the congress, see Lehto, *Mathematics without Borders*, 24–33; and Curbera, *Mathematicians of the World, Unite!*, 69–74. Also the substantial volume of proceedings: *Comptes rendus du Congrès international des mathématicens (Strasbourg, 22–30 septembre 1920) : Publiés par Henri Villat* (Toulouse, 1921).

28 Mittag-Leffler to Eliakim, 8 March 1921, quoted in Lehto, *Mathematics without Borders*, 24.

29 Quoted as "pardonner à certains crimes, c'est s'en faire le complice, " in "Allocution de M. Emile Picard," in *Comptes rendus du Congrès international des mathématiciens . . . Strasbourg, 1920*, xxxi–xxxiii (esp. xxxiii).

30 On the foundation of the IMU and the proceedings of the first General Assembly, see the report by Gabriel Koenigs, as secretary-general of the congress,

in *Comptes rendus du Congrès international des mathématiciens . . . Strasbourg, 1920*, xxxiv–xxxix (esp. xxxiv–xxxvi).

31 See chapter 2, 63–64.

32 On Strasbourg and its university in this period, see John Craig, *Scholarship and Nation Building: The Universities of Strasbourg and Alsatian Society, 1879–1939* (Chicago and London, 1984), 5–194; and the essays in Elisabeth Crawford and Olff-Nathan, eds., *La Science sous influence : L'université de Strasbourg, enjeu des conflits franco-allemands 1872–1945* (Strasbourg, 2006), especially Elisabeth Crawford's introduction, "L'histoire des universités de Strasbourg: état des lieux," 7–12; John Craig, "La Kaiser-Wilhelms-Universität Strassburg 1872–1918," 15–28; and Christian Bonah, "Une université internationale malgré elle," 29–35.

33 David Cahan, "The Institutional Revolution in German physics, 1865–1914," *Historical Studies in the Physical Sciences* 15, part 2 (1985): 1–65 (esp. 25–33) endorses contemporary judgements in presenting the Strasbourg Physics Institute as "the model institute."

34 Françoise Olivier-Utard, "L'Université de Strasbourg: un double défi, face à l'Allemagne et face à la France," in Crawford and Olff-Nathan, *Science sous influence*, 137–172.

35 Christian Pfister, " L'Université de Strasbourg," *Revue politique et littéraire. Revue bleue* 59e année (17 December 1921): 753–760 (esp. 758–760).

36 Louis Lumet, *Pasteur : Sa vie, son oeuvre* (Paris, 1922).

37 *République française. Ministère du travail, de l'hygiène, de l'assistance et de la prévoyance sociales : Livre d'or de la commémoration nationale du centenaire de la naissance de Pasteur célébrée du 24 au 31 mai 1923* (Paris, 1928), esp. 250–306, on the events in Strasbourg. See also the volume marking the participation of the city of Paris in the celebrations: René Weiss, *La Commémoration du centenaire de Pasteur par la ville de Paris* (Paris, 1924), esp. 119–193, on the city's representation in Strasbourg. A brief contemporary report is Maurice Deschiens, "L'apothéose de Pasteur: les fêtes nationales du centenaire de la naissance de Pasteur," *Chimie & industrie* 9, no. 6 (June 1923): 1057–1075.

38 The Institut d'hygiène et de bactériologie was founded in 1919 as the second of the two Instituts Pasteur in provincial France, following the IP in Lille (1894).

39 On the exhibition, see the 246-page official catalogue, *République française. Ministère de l'Hygiène, de l'Assistance et de la Prévoyance sociales : Hygiène scientifique et appliquée. Catalogue officiel* (n. p., 1923).

40 Millerand's main speech in Strasbourg pulled no punches; see *Livre d'or de la commémoration nationale*, 251–253.

41 Augustin Boutaric, *Marcellin Berthelot (1827–1907)* (Paris, 1927).

42 On the centenary, see Robert Fox, "Science, Celebrity, Diplomacy: The Marcellin Berthelot Centenary, 1927," *Revue d'histoire des sciences* 69 (2016): 77–115; p. 83 on the Pasteur and Berthelot stamps. The week's events are recorded, with illustrations and reproduced documents, in the sumptuous commemorative volume *Centenaire de Marcelin Berthelot, 1827–1927* (Paris, 1929).

43   On the plans for the Maison de la Chimie, see Danielle M. E. Fauque, "French Chemists and the International Reorganisation of Chemistry after World War I," *Ambix* 58 (2011): 116–135; and "La documentation au cœur de la réorganisation de la chimie dans l'entre-deux- guerres: rôle des sociétés savantes et institutions françaises dans le contexte international," *Revue d'histoire des sciences*, 69 (2016): 41–75.

44   See chapter 1, 33.

45   Despite the additional difficulties of the occupation of Paris during the Second World War, the library and related services of the Maison de la chimie continued to function, still in their original premises, until the library ceased collecting and was transferred in 1964 as a nucleus for the library of the new university at Orsay, now the Université Paris-Sud XI. The library, recently recatalogued under the direction of François Léger, is kept as a special collection.

46   "Master work" in the sense enshrined in the museum's official title: Deutsches Museum von Meisterwerken der Naturwissenschaft und Technik. On the conception and early development of the Deutsches Museum, see Maria Osietzki, "Die Gründungsgeschichte des Deutschen Museums von Meisterwerken der Naturwissenschaften und Technik in München 1903–1906," *Technikgeschichte* 52 (1985): 49–75; also Wolfhard Weber, "Vorgeschichte und Voraussetzungen der Museumsgründung," and Wilhelm Füßl, "Gründung und Aufbau 1903–1925," both in Füßl and Helmuth Trischler, eds., *Geschichte des Deutschen Museum: Akteure, Artefakte, Ausstellungen* (Munich, 2003), 45–58 and 59–101.

47   The origins of what was initially called the Technical Museum of the Czech Kingdom are described in Josef Gruber, *Technické museum pro království České* (Prague, 1908). I am grateful to Marcela Efmertova for drawing my attention to this work. Although the Technical Museum in Vienna had roots going back to the early nineteenth century, the main source of its collections was a variety of private and state technical museums, given coherence in the early years of the twentieth century by the vision of an academic professor of mechanics, Wilhelm Franz Exner. Fine essays on the origins and eventual opening of the museum (in 1918) are in Helmut Lackner, Katharina Jesswein, and Gabriele Zuna-Kratky, eds., *100 Jahre Technichsche Museum Wien* (Vienna, 2009). On the origins of the Science Museum collections, see Robert Bud, "Infected by the Bacillus of Science: The Explosion of South Kensington," in Peter J. T. Morris, ed., *Science for the Nation: Perspectives on the History of the Science Museum* (Basingstoke, 2010), 11–40, and other essays in the same volume.

48   Svante Lindqvist, "An Olympic Stadium of Technology: Deutsches Museum and Sweden's Tekniska Museet," in Brigitte Schroeder-Gudehus, ed., *Industrial Society and its Museums 1890–1990: Social Aspirations and Cultural Politics* (Chur, 1993), 37–54 (esp. 38).

49   A point well made by Lindqvist; ibid., 45–50.

50   Tom Scheinfeldt, "Sites of Salvage: Science History between the Wars" (University of Oxford, D.Phil. thesis, 2003), 191–203.

51 Alan Q. Morton, "The Electron Made Public: The Exhibition of Pure Science in the British Empire Exhibition, 1924–5," in Bernard Finn, Robert Bud, and Helmuth Trischler, eds., *Exposing Electronics* (Artefacts: Studies of the History of Science and Technology, vol. 2; London, 2003), 25–43 (esp. 27–29 and 37). The words quoted appear in the descriptive catalogue of the exhibits, published with a set of thematic essays stressing British achievements, in *British Empire Exhibition, 1924: Handbook to the Exhibition of Pure Science. Arranged by the Royal Society* (London, 1924), 145. Morton contrasts the tone of the Exhibition of Pure Science with the explicitly international character of the Special Loan Exhibition of Scientific Apparatus mounted in the South Kensington Museum in 1876. He also notes the essentially domestic concern with the status of physicists, promoted by the Institute of Physics, founded in 1921.

52 Ben Russell, "Watt's Workshop: Craft and Philosophy in the Science Museum," *Science Museum Group Journal*, no. 1 (spring 2014). DOI: http://dx/doi.org/10.15180/140106. For an illuminating discussion of Watt's status as a particular kind of British hero, see Christine MacLeod, *Heroes of Invention: Technology, Liberalism and British Identity, 1750–1914* (Cambridge, 2007), 91–124.

53 See http://www.museogalileo.it/istituto/imuseonazionalestoriascienze1930 1945.html for a brief account. The opening of the museum followed the previous year's successful Esposizione Nazionale di Storia della Scienza in Florence. Giuseppe Boffito, *Gli Strumenti della scienza e la scienza degli strumenti : Con l'illustrazione della Tribuna di Galileo* (Florence, 1929), published on the occasion of the 1929 exhibition, contains a separately paginated inventory of the instruments on show, an account of the origins of the collection, and a pioneering essay on the history of scientific instruments.

54 Eva A. Mayring, "Das Porträt als Programm," in Ulf Hashagen, Oskar Blumtritt, and Helmuth Trischler, eds., *Circa 1903: Artefakte in der Gründungszeit des Deutschen Museum* (Munich, 2003), 54–77, where some of the early portraits are reproduced.

55 The challenges and compromises of these years are treated in detail in the contributions to Elisabeth Vaupel and Stefan L. Wolff, eds., *Das Deutsche Museum in der Zeit des Nationalsozialismus : Eine Bestandsaufnahme* (Göttingen, 2010). For helpful overviews, see Eve M. Duffy, "Jenseits von Anpassung und Autonomie: zur institutionellen Entwicklung des Deutschen Museums zwischen 1933 und 1945," ibid., 45–77; and the same author's "Im Spannungsfeld von Selbststeuerung und Fremdbestimmung 1925–1944," in Füßl and Trischler, *Geschichte des Deutschen Museum*, 103–147.

56 Frank Uekötter, "Expansionsgelüste an der Isar. Das Deutche Museum und die Führung des Dritten Reichs: Adolf Hitler, Fritz Todt und die Pläne für ein Haus der deutschen Technik," in Vaupel and Wolff, *Das Deutsche Museum in der Zeit des Nationalsozialismus*, 195–243.

57 Stefan L. Wolff, "Jonathan Zenneck als Vorstand des Deutschen Museums," ibid., 78–126 (107).

58 When the exhibition moved on to Berlin in 1937, the title was changed to "Bolschevismus ohne Maske" ("Bolshevism unmasked").

59 The two exhibitions are discussed in Wolfgang Benz, *"Der ewige Jude": Metaphern und Methoden nationalsozialistischer Propaganda* (Berlin, 2010), 47–73; and the same author's "Die Ausstellung 'Der ewige Jude,'" in Vaupel and Wolff, *Das Deutsche Museum in der Zeit des Nationalsozialismus*, 652–680.

60 See Benz, *"Der ewige Jude,"* 97–114; and "Die Ausstellung 'Der ewige Jude,'" 659, for images of this part of the exhibition.

61 On the Nazi concept of "degenerate art" and the 1937 exhibition, see Olaf Peters, ed., *Degenerate Art: The Attack on Modern Art in Nazi Germany, 1937* (Munich, London, and New York, 2014), a finely illustrated catalogue of the exhibition on the subject in New York in 2014. A helpful, briefer account is Shearer West, *The Visual Arts in Germany 1890–1937: Utopia and Despair* (Manchester, 2000), 189–193.

62 Philipp Lenard, *Deutsche Physik in vier Bänden*, 4 vols. (Munich, 1936–1937), vol. I, ix–xv (ix–xi).

63 On all aspects of Volterra's life, see Angelo Guerraggio and Giovanni Paoloni, *Vito Volterra* (Heidelberg, etc., 2013).

64 Ibid., 82–109; Giuliano Pancaldi, "Vito Volterra: Cosmopolitan Ideals and Nationality in the Italian Scientific Community between the *Belle Epoque* and the First World War," *Minerva* 31 (1993): 21–37 (esp. 32–35); and Raffaella Simili and Giovanni Paoloni, eds., *Per una storia del Consiglio nazionale delle ricerche*, 2 vols. (Rome, 2001), esp. Luigi Tomassini, "Le origini," and Simili, "La presidenza Volterra," vol. I, 5–71 and 72–127.

65 On the persistent harassment of Volterra in his later years, see Gueraggio and Paoloni, *Vito Volterra*, 122–131 and 153–156. The letters from the rector of the University of Rome presenting the terms of the oath and, on December 31, 1931, confirming Volterra's enforced retirement ("per incompatibilità con le generali direttivi politiche del Governo") are reproduced in Giovanni Paoloni, ed., *Vito Volterra e il suo tempo (1860–1940): Mostra storico-documentaria* (Rome, 1990), figures VII.8 and VII.9. Volterra served as an honorary president of the IMU from its foundation in 1920 until its suspension in 1932.

66 José Manuel Sánchez-Ron and Antoni Roca-Rosell, "Spain's First School of Physics: The Laboratorio de Investigaciones Físicas," *Osiris* 8 (1993): 127–155.

67 In an extensive secondary literature on the history of the JAE, a good overview is José Manuel Sánchez-Ron, "La Junta para Ampliación e Investigaciones Científicas (1907–2007)," in *El Laboratorio de España: La Junta para Ampliación de Estudios e Investigaciones Científicas 1907–2007* (Madrid, 2007), 64–125. This same collective volume contains other valuable studies, as does José Manuel Sánchez-Ron and José García-Velasco, eds., *100 JAE. La Junta para Ampliación de Estudios e Investigaciones Científicas en su Centenario : Actas del II Congreso Internacional, celebrado los días 4, 5 y 6 de Febrero de 2008*, 2 vols. (Madrid, 2010), a collection of papers arising from the international colloquium held in Madrid in 2008 to mark the JAE's centenary.

68 Isabel Pérez-Villanueva Tovar, "La Residencia de Estudiantes," in *El Laboratorio de España*, 432–463.

69 The special importance of visitors to the Residencia in the crucial years for physics between 1923 and 1932 emerges strongly from the essays in José Manuel Sánchez Ron, ed., *Creadores científicos: la física en la Residencia de Estudiantes* (Madrid, 2013).

70 Giovanni Paoloni, "Marconi, la politica e le istituzioni scientifiche italiane negli anni trenta," in *Cento anni di radio : Da Marconi al futuro delle telecommunicazioni* (Venice, 1995), 35–37; and Giovanni Paoloni and Raffaella Simili, "Scienza, impresa, amministrazione: Marconi e le istituzioni italiane," in Gabriele Falciasecca and Barbara Valotti, eds., *Guglielmo Marconi : Genio, Storia e Modernità* (Milan, 2003), 97–111. Giancarlo Masini, *Guglielmo Marconi* (Turin, 1975), 436–441, offers an interesting account of the funeral. The Accademia was established in 1926 with a status that set it apart from the other, more cosmopolitan Italian academies. Its first members were chosen in 1929, but the Accademia barely functioned until Marconi's appointment in the following year.

71 On Marconi's status as "Un idolo del regime" and his growing disillusionment, see Masini, *Guglielmo Marconi*, 366–411; also, briefly but interestingly, Degna Marconi, *My Father, Marconi* (London, 1962), 271–273. The effect of the racial laws on Italian academic life is discussed in essays by Annalisa Capristo and Roberto Finzi in Joshua Z. Zimmerman, ed., *Jews in Italy under Fascist and Nazi Rule, 1922–1945* (Cambridge, 2005), 81–133.

72 Marconi, *My Father, Marconi*, 292–293.

73 Loren R. Graham, *The Soviet Academy of Sciences and the Communist Party, 1927–1932* (Princeton, NJ, 1967), esp. 80–153; and *Science and Philosophy in the Soviet Union* (New York, 1971), 3–23, offer helpful introductions to the issues.

74 David Joravsky, *The Lysenko Affair* (Cambridge, MA, 1970), esp. 63–96; and Graham, *Science and Philosophy in the Soviet Union*, 195–256.

75 Jacqueline Eidelman, "The Cathedral of French Science: The Early Years of the 'Palais de la découverte,'" in Terry Shinn and Richard Whitley, eds, *Expository Science: Forms and Functions of Popularisation* (Dordrecht, Boston, and Lancaster, 1985), 195–207.

76 Charlotte Bigg and Andrée Bergeron, "D'ombres et de lumières: l'exposition de 1937 et les premières années du Palais de la découverte au prisme du transnational," *Revue germanique internationale* 21 (2015): 187–206. In their discussion of the internationalism of Perrin, Langevin, and Joliot that informed the 1937 Exposition, Bigg and Bergeron explore a dimension that complements Eidelman, "The Cathedral of French Science."

77 For a detailed description of the Palais de la Découverte, see the official report on the exhibition: *Ministère du Commerce et de l'Industrie: Exposition internationale des arts et techniques. Paris 1937. Rapport général*, 11 vols. (Paris, 1938–1940), vol. 4, 215–565. Briefer accounts are in Paul Dupays, *L'Exposition internationale de 1937 : Ses créations et ses merveilles* (Paris, 1938), 60–70 ; and the official English-language guidebook, *International Exhibition. Paris 1937: Arts*

and Crafts in Modern Life Official Guide (Paris, 1937), 74–80.

78 The judgement appears in the report cited in the previous note: *Rapport général*, vol. 4, 224.

79 Karen Fiss, *Grand Illusion: The Third Reich, the Paris Exposition, and the Cultural Seduction of France* (Chicago, 2009), esp. 44–128; and Robert H. Kargon, Karen Fiss, Morris Low, and Arthur P. Molella, *World's Fairs on the Eve of War: Science, Technology & Modernity, 1937–1942* (Pittsburgh, 2015), 7–29.

80 Fiss, *Grand Illusion*, 59–62; and, by the same author, "In Hitler's Salon: The German Pavilion at the 1937 Paris Exposition Internationale," in Richard A. Eitlin, ed., *Art, Culture, and Media under the Third Reich* (Chicago and London, 2002), 316–342.

81 Kargon et al., *World's Fairs on the Eve of War*, 108–137. For other rich sources on the exhibition, see the two volumes of *E42: Utopia e scenario del regime* (Venice, 1987), edited by Tullio Gregory and Achille Tartaro (vol. 1) and Maurizio Calvesi, Enrico Guidoni, and Simonetta Lux (vol. 2) and, for a fine account of the exhibition's links with contemporary cultural policy in Nazi Germany, Nicola Timmermann, *Repräsentative "Staatsbaukunst" im faschistischen Italien und im nationalsozialistischen Deutschland—der Einfluß der Berlin-Planung auf die EUR* (Stuttgart, 2001), 111–263.

82 "una rappresentazione delle attività creatrici dello spirito umano, ovunque e comunque si siano menifestate, nel dinamismo di quella perenne e feconda emulazione, in cui è la garanzia del progresso universale, dagli illimitati termini e confine," in, *Esposizione Universale di Roma. MCMXLII, Anno XXo E.F. (1939). Rome: A Cura del Commissariato Generale nell'anno XVIIo* (Livorno, 1942), 21.

83 Ibid., 56–60.

84 Paolo Galluzzi, "La storia della scienza nell'E42," in Gregory and Tartaro, *Utopia e scenario del regime. Vol. 2. Ideologia e programma per l'"Olimpiade delle civiltà,"* 53–69; and Kargon et al., *World's Fairs on the Eve of War*, 128–134.

85 H. G. Wells, *World Brain* (London, 1938). For a reflective study of Wells's concept, see W. Boyd Rayward, "H. G. Wells's Idea of a World Brain: A Critical Re-assessment," *Journal of the American Society for Information Science* 50 (1999): 557–579.

86 Wells, "World Encyclopaedia" (a lecture given at the Royal Institution on November 20, 1936) and "The Idea of a Permanent World Encyclopaedia," both in Wells, *World Brain*, 1–25 and 58–62. The essay "The Idea of a Permanent World Encyclopaedia," dated August 1937, was originally written for the *Encyclopédie française*, where it appeared as "Rêverie sur un thème encyclopédique," in *Encyclopédie francaise*, Lucien Febvre and Gaston Berger, eds., 21 vols. (Paris, 1935–1966), vol. 18 ("La civilisation écrite," ed. Julien Cain, July 1939), 18·24—11–12.

87 H. G. Wells, *The Outline of History: Being a Plain History of Life and Mankind*, originally published in twenty-four fortnightly parts, beginning in November 1919. By the time of the fourth, "definitive" edition (London, 1923), the work was arranged in forty-one chapters.

88 "De tout temps la science a été une oeuvre internationale" and "Le machinisme doit ses progrès à tous les travailleurs du monde." Both statements are from the illustrated brochure incorporating visual material presented by the Commission française de coopération intellectuelle, with a preface by Edouard Herriot advancing intellectual exchanges as a vehicle for peace: *Coopération intellectuelle: Les échanges intellectuelles à travers le monde. Paris 1937* (Bod. 3975 c. 6).

89 The words quoted are from the Introduction (by Paul Valéry and Henri Focillon) to the "open letters," by leading figures in the ICIC, collected in Focillon et al., *A League of Minds* (Paris, 1933), 11–21.

Epilogue

1 Jean-Jacques Renoliet, *L'UNESCO oubliée : La Société des Nations et la coopération intellectuelle (1919–1946)* (Paris, 1999), 151–178.

2 Ibid., 158. On the effect of the Second World War on the ICIC and intellectual cooperation more generally, see also Corinne A. Pernet, "Twists, Turns and Dead Alleys: The League of Nations and Intellectual Cooperation in Times of War," *Journal of Modern European History* 12 (2014): 342–358.

3 Joseph and Dorothy Needham, *Science Outpost: Papers of the Sino-British Science Co-operation Office (British Council Scientific Office in China) 1942–1946* (London, 1948).

4 The proceedings of the conference were published in a special issue of the BAAS journal, *The Advancement of Science* 2, no. 5 (1942). A "Penguin Special" paperback based on the proceedings appeared as J. G. Crowther, O. J. R. Howarth, and D. P. Riley, *Science and World Order* (Harmondsworth and New York, 1942).

5 Ibid., 115–116, for the "Declaration" and Gregory's presentation of it.

6 *Advancement of Science* 2, no. 5 (1942): "Introduction," 3. The words paraphrased the thrust of a speech by the British foreign minister, Anthony Eden, on the eve of the conference.

7 Renoliet, *L'UNESCO oubliée*, 161–178.

8 Jean-Noël Jeanneney, *Google and the Myth of Universal Knowledge: A View from Europe*, trans. Teresa Lavender Fagan (Chicago and London, 2007).

9 On the difficulties of the Google project, see Robert Darnton, "Google's Loss: The Public's Gain," *New York Review of Books*, April 28, 2011.

10 The Europeana homepage at http://www.europeana.eu/portal/ gives access to the sites and catalogues of more than two thousand contributing libraries, museums, archival depositories, and other institutions.

11 Robert Darnton, "The National Digital Public Library Is Launched," *New York Review of Books*, April 25, 2013. The DPLA's homepage is at http://dp.la/.

12 Francis Heylighen, "Das globale Gehirn als neues Utopia," in Rudolf Maresch and Florian Rötzer, eds., *Renaissance der Utopie. Zukunftsfiguren des 21. Jahrhunderts* (Frankfurt, 2004), 92–112.

13 As Anita Guerrini has suggested to me, contemporary critics who might have sympathized with Wells's humanitarian concerns would also have been uneasy about the authoritarian tone of the "world-socialism" that he professed in his controversial book, *The New World Order: Whether It Is Attainable, How It Can Be Attained, and What Sort of World a World of Peace Will Have to Be* (London, 1940).

14 From the home-page of the Global Brain Institute at https://sites.google.com/site/gbialternative1/.

BIBLIOGRAPHICAL ESSAY

Historians working in the broad areas of universalism, cosmopolitanism, transnationalism, internationalism, and their variants and subsets in the nineteenth and twentieth centuries can draw on a secondary literature that has grown significantly in recent years. Setting the bounds for an appropriate bibliography is no easy matter, and in this essay I do no more than highlight a small selection of what I see as the key relevant sources. In my choice of items to mention, I have tended to favor those with substantial bibliographical elements as a guide to further reading.

On all aspects of internationalism in science, Brigitte Schroeder-Gudehus has set a formidably high standard since her pioneering work began to appear in the 1960s. Following her book *Les Scientifiques et la paix: La Communauté scientifique internationale au cours des années 20* (Montreal: Presses universitaires de Montréal, 1978), she has published extensively on the institutions of international science and on the place of exhibitions and museums in modern industrial culture. Her collaboration with Anne Rasmussen resulted in an essential guide to universal exhibitions since 1851: *Les Fastes du progrès: Le guide des expositions universelles, 1851–1992* (Paris: Flammarion, 1992), with helpful editorial material and a rich body of bibliographical and other information. On museums, *La Société industrielle et ses musées: Demande sociale et choix politiques, 1890–1990* (Paris: Editions des Archives contemporaines, 1992), edited by Schroeder-Gudehus with Rasmussen and Edgar Bolenz, is an important collection of essays, a selection of which subsequently appeared in an English edition as *Industrial Society and Its Museums 1890–1990: Social Aspirations and Cultural Politics* (Chur: Harwood Academic, 1993); also published, with the same pagination, as a special issue of *History and Technology: An International Journal* 10, no. 1 (1993).

Rasmussen herself has broken other important new ground. Her doctoral thesis, "L'Internationale scientifique 1890–1914" (Ecole des Hautes Etudes en Sciences Sociales, Paris, 1995) is a fine wide-ranging study of pre–First World War internationalism in science; the copy I have consulted is in the Bibliothèque des sciences et de l'industrie, Cité des sciences et de l'industrie, La Villette, Paris. Her "Jalons pour une histoire des congrès internationaux au XIXe siècle: régulation scientifique et propagande intellectuelle" appeared in an important thematic issue of *Relations internationales* (no. 62, 1990, 115–133) devoted to international scientific congresses. Schroeder-Gudehus's "Les congrès scientifiques et la politique de coopération internationale des académies des sciences," in the same issue (135–148), integrates a discussion of congresses with the interactions between academies exemplified in the work of the International Association of Academies. Scientific congresses have also been the subject of a special issue of *Revue germanique internationale* (no. 12, 2010) on "La fabrique internationale de la science. Les congrès scientifiques de 1865 à 1945," though with a coverage extending beyond congresses of a strictly scientific nature.

Research in areas of such diversity invites collaboration, and it is no coincidence that some of the best recent work is to be found in edited volumes that open up comparative perspectives transcending individual countries and cultures. Notable among such volumes are Madeleine Herren, *Hintertüren zur Macht: Internationalismus und modernisierungsorientierte Außenpolitik in Belgien, der Schweiz und den USA 1865–1914* (Munich: Oldenbourg, 2000); and Martin H. Geyer and Johannes Paulmann, eds., *The Mechanics of Internationalism: Culture, Society, and Politics from the 1840s to the First World War* (Oxford: German Historical Institute and Oxford University Press, 2001).

In chapter 1 of *Science without Frontiers*, I explore the interface between bibliographical initiatives and broader movements aimed at extending understanding and hence peace between nations. In treating that subject, I have benefited from a secondary literature that has grown dramatically since W. Boyd Rayward published *The Universe of Information: The Work of Paul Otlet for Documentation and International Organisation* (Moscow: Published for the International Federation for Documentation [FID] by All-Union Institute for Scientific and Technical Information [VINITI], 1975). Rayward's book, which is now available online on "Rayward's Otlet page," is a rich study of Otlet informed by the author's discovery (in the late 1960s) of what remained in Brussels of the Mundaneum (http://people.lis.illinois.edu/~wrayward/otlet/otletpage.htm). Via his homepage, Rayward has made available not only a number of his own publications but also valuable visual material, including a half-hour film concerning his first encounters with the Otlet legacy (https://archive.org/details/paulotlet).

After several decades of neglect, Otlet's contributions to bibliography and library science have now been recognized. This has resulted in the installation of a visitors' center and archive, in a former department store in Mons, devoted to the history of information management and retrieval. Inheriting the name Mundaneum, the display and research collection (under the head of the archives, Stéphanie Manfroid) offer a fascinating insight into Otlet's achievements and their legacy. Recent books that draw on the materials in Mons include two important biographies that build on Rayward's pioneering work: Françoise Levie's finely illustrated *L'Homme qui voulait classer le monde: Paul Otlet et la Mundaneum* (Brussels: Les Impressions nouvelles, 2006) and Alex Wright's *Cataloging the World: Paul Otlet and the Birth of the Information Age* (New York: Oxford University Press, 2014).

As I argue in chapters 1 and 2, the history of the Nobel Prizes offers rich insights into the world of international science, and the subject is served by a plentiful literature, to which Elisabeth Crawford and Robert Friedman have made particularly significant contributions. For Crawford's work, see her two books: *The Beginnings of the Nobel Institution: The Science Prizes, 1901–1915* (Cambridge: Cambridge University Press and Paris: Editions de la Maison des sciences de l'homme, 1984), and *Nationalism and Internationalism in Science, 1880–1939: Four Studies of the Nobel Population* (Cambridge: Cambridge University Press, 1992). Among Friedman's publications, see especially *The Politics of*

Excellence: Behind the Nobel Prize in Science (New York: Times Books, 2012). On the pattern of nominations for the various prizes, a rich database, "Nomination Archive" is now available on the Nobel Prize website (nobelprize.org/nomination/archive). For the prizes in physics and chemistry, an alternative source, with a helpful introductory essay, is Elisabeth Crawford, J. L. Heilbron, and Rebecca Ulrich, *The Nobel Population 1901–1937: A Census of the Nominators and Nominees for the Prizes in Physics and Chemistry* (Berkeley, CA: Office for the History of Science and Technology, University of California, Berkeley; and Office for History of Science, Uppsala University, Uppsala, 1987).

In different ways, most of the sources I have mentioned throw light on a pre-1914 belief in the beneficence of science and the free exchange of knowledge that it is hard to recover from our twenty-first-century perspective, clouded as it is by a history of two world wars and a cold war. One internationally minded figure whose work for peace encapsulated the belief and whose utopian ideas and eventual disillusionment I discuss in chapters 1 and 2 was the Norwegian-American artist Hendrik Christian Andersen. The privately printed *Creation of a World Centre of Communication by Hendrik Christian Andersen: Ernest M. Hébrard Architect* (Paris, 1913) and the supplementary volume that Andersen published with his sister-in-law Olivia Cushing Andersen, also entitled *Creation of a World Centre of Communication* (Rome, 1918), convey the grandeur of his vision and are among the most sumptuously produced twentieth-century books it has been my pleasure to handle.

Among other dedicated internationalists, it is encouraging to see Wilhelm Ostwald, Nobel Prize winner for chemistry in 1909, attracting new attention, for example in Martin Kohlrausch and Helmuth Trischler, *Building Europe on Expertise: Innovators, Organizers, Networkers* (Basingstoke: Palgrave Macmillan, 2014), esp. 102–107 (in a chapter titled "Architectures of knowledge" that also treats the work of Paul Otlet and Otto Neurath). On Ostwald's campaign in support of the artificial language Ido (as on all aspects of the relations between science and language since the retreat of Latin), see Michael Gordin's groundbreaking *Scientific Babel: The Language of Science from the Fall of Latin to the Rise of English* (London: Profile Books, 2014), esp. 131–185, where Gordin also examines the far-reaching linguistic consequences of the First World War. Roswitha Reinbothe's *Deutsch als internationale Wissenschaftssprache und der Boykott nach dem Ersten Weltkrieg* (Frankfurt am Main: Peter Lang, 2006) focuses specifically on the postwar ban on the use of German in international meetings. For an outline of her argument, in French, see "L'exclusion des scientifiques allemands et de la langue allemande des congrès scientifiques internationaux après la Première Guerre mondiale," in "La fabrique internationale de la science," the special issue (mentioned above) of the *Revue germanique internationale* 12 (2010), 193–208.

In presenting the 1914–1918 war as a watershed between early-twentieth-century cosmopolitan ideals and the nationally driven competitiveness of the interwar years, I have not felt it necessary to dwell at any length on the contributions of scientists and engineers whose expertise was mobilized. But what I say about war-related science in chapter 2 rests on a literature still growing rapidly through conferences and other initiatives to mark the centenary of what has come to be conventionally known as the chemists' war. The use of poison gases, first deployed on the western front near Ypres in April 1915, was a new departure that transformed public perceptions of the role of science in the conflict. Here, L. F. Haber, *The Poisonous Cloud: Chemical Warfare in the First World War* (Oxford: Clarendon Press, 1986), and Olivier Lepick, *La Grande Guerre chimique : 1914–1918* (Paris: Presses universitaires de France, 1998), are fine, complementary sources. On the contribution of chemists and the chemical industry to the British war effort more generally, see Roy M. MacLeod, "The Chemists Go to War: The Mobilization of Civilian Chemists and the British War Effort, 1914–1918," *Annals of Science* 50 (1993), 455–481. For a perspective extending beyond Britain, see also MacLeod's "The Scientists Go to War: Revisiting Precept and Practice, 1914–1919," *Journal of War and Culture Studies* 2 (2009), 37–51, and the collection of essays that he has edited with Jeffrey A. Johnson: *Frontline and Factory: Comparative Perspectives on the Chemical Industry at War, 1914–1924* (Dordrecht: Springer, 2004). MacLeod is currently assembling a collective volume, *The Great War and Modern Science* (Palgrave Macmillan, forthcoming).

The recent growth of interest in the reordering of international science after the First World War is striking. On relations between the scientific communities of the former belligerent nations, however, Frank Greenaway, *Science International: A History of the International Council of Scientific Unions* (Cambridge: Cambridge University Press, 1996) remains a standard source; there is no better way into the history of the International Research Council and its transformation, in 1931, into the International Council of Scientific Unions. For a briefer introduction, however, see also Daniel J. Kevles, "Into Hostile Camps: The Reorganization of International Science after World War I," *Isis* 62 (1971), 47–69. On the same theme, Anne Rasmussen's essay "Réparer, réconcilier, oublier: enjeux et mythes de la démobilisation scientifique, 1918–1925," *Histoire@Politique. Politique, culture, société*, no. 3 (November-December 2007), analyses the resistance of many scientists to the process of "scientific demobilization" and the consequent perpetuation into the mid-1920s of enmities bred of the war (http://www.histoire-politique.fr/index.php?numero=03). Of individual scientific disciplines, chemistry has attracted particular attention, for example in "Chemistry in the Aftermath of World Wars," a special issue of *Ambix* (58, no. 2, July 2011) that includes important essays, among others, on the role of French and Japanese chemists in the reorganization of international chemistry after 1918, as well as an overview by the issue's guest editor, Jeffrey Johnson. Physics, however, has not been neglected. Here, Paul Forman's "Scientific Internationalism and the Weimar Physicists: The Ideology and Its Manipulation in Germany after World War I," *Isis* 64 (1973), 150–180 remains a classic study of the complex attitudes to internationalism in the Weimar physics community. Inter-war physics in Germany is also the subject of Andreas Kleinert, "Von der Science allemande zur deutschen Physik: Nationalismus und moderne Naturwissenschaft in Frankreich und Deutschland zwischen 1914 und 1940," *Francia* 6 (1978), 509–525, a study of the origins of the notion of a truly German physics, as promulgated by Philipp Lenard and Johannes Stark as part of National Socialist orthodoxy.

While the former Allies and Central powers remain best-studied, the trend to a wider geographical focus is evident in a collection of studies of the response of neutral nations to the readjustments in international science after 1918: Rebecka Lettevall, Geert Somsen, and Sven Widmalm, eds., *Neutrality in Twentieth-Century Europe: Intersections of Science, Culture, and Politics after the First World War* (New York and London: Routledge, 2012) contains essays on Sweden, Holland, Switzerland, Norway, Czechoslovakia, and Austria, as well as more general contributions, including a helpful editors' introduction and a reflection on the concepts of internationalism and neutrality in science by Brigitte Schroeder-Gudehus. As the volume shows, consideration of the noncombatant nations is essential to any understanding of the impact of the First World War on interwar science. A good illustration of the opportunities for further research in this area is Jorrit Amit, "Nuclei in a Supersaturated Solution: Utrecht Chemists and the Crystallization of International Relations after the First World War," *Studium: Tijdschrift voor Wetesnchaps- en Universiteitsgeschiedenis. Revue d'histoire des sciences et des universités* 7 (2014), 190–208, a discipline-focused study of the context and consequences of the "International Chemical Reunion" conference, held in Utrecht in 1922. Largely the work of the internationally minded Dutch chemist Ernst Cohen, the conference brought together representatives from Germany, Britain, the United States, and Holland in an early gesture of reconciliation, though in the absence of any French or Belgian involvement.

While internationalist sentiment and national interests are useful analytical categories, there is a danger in seeing them as clearly defined and systematically antagonistic forces. Glenda Sluga's *Internationalism in the Age of Nationalism* (Philadelphia: University of Pennsylvania Press, 2013) shows how intertwined the two have been and remain in our own day. For Sluga, nationalism and internationalism are best viewed as existing in an "oscillating relationship (6)." Focusing on the specific context of Belgium, Daniel Laqua, *The Age of Internationalism and Belgium, 1880–1930* (Manchester and New York: Manchester University Press, 2013) shows how Belgian national aspirations and a governmentally promoted internationalism worked together, most strongly before 1914 though also on into the 1920s. Such a pattern of coexistence is a recurring theme of *Science without Frontiers* as well. Hence, in stressing the transnational tone of science from the mid-nineteenth century to the First World War, as I do in chapter 1, I do not wish to imply that national interests were absent in this period. The streak of competitiveness that characterized universal exhibitions and most multinational collaborative ventures in science gave rise to what Geert J. Somsen has analysed as an "Olympic" form of internationalism that made success in an exhibition or the debates at a scientific congress a matter of national pride; see "A History of Universalism: Conceptions of the Internationality of Science from the Enlightenment to the Cold War," *Minerva* 46 (2008), 361–379, in which Somsen broaches issues relevant to the whole notion of universalism. The essays in Mitchell G. Ash and Jan Surman, eds., *The Nationalization of Scientific Knowledge in the Habsburg Empire, 1848–1918* (Basingstoke: Palgrave Macmillan, 2012) go further in exploring the role of science in nation-building and hence in the undermining of the once broadly unified and predominantly German-speaking central European "republic of letters." Studies of Austria, Poland, the Czech lands, and Hungary converge in the volume to a pattern of fragmentation carried forward by scientists sensitive to their local national allegiances and languages as well as to contemporary perceptions of science as an intrinsically international culture.

By the same token, as I argue in chapter 3, scientific internationalism lived on in the interwar years in a world increasingly marked by national priorities. It found an eloquent voice in the International Commmittee on Intellectual Co-operation. Jean-Jacques Renoliet, *L'UNESCO oubliée: La Société des Nations et la coopération intellectuelle (1919–1946)* (Paris: Publications de la Sorbonne, 1999) brings out well both the ICIC's successes and its limitations as the modest, though intellectually distinguished, cultural wing of the League of Nations. The fact remains, however, that trends that worked against internationalist sentiment between the two world wars were more powerful. Here, histories of national museums offer valuable insights. Helmut Lackner, Katharina Jesswein, and Gabriele

Zuna-Kratky, eds., *100 Jahre Technisches Museum Wien* (Vienna: Verlag Carl Uberreuter, 2009) (a richly illustrated volume published to mark the centenary of the founding of the Technical Museum in Vienna) and Peter J. M. Morris, ed., *Science for the Nation: Perspectives on the History of the Science Museum* (London: Palgrave Macmillan and Science Museum, 2010) are notable examples. A volume that explores a particularly vivid instance of the "national turn" in museums is Elisabeth Vaupel and Stefan L. Wolff, eds., *Das Deutsche Museum in der Zeit des Nationalsozialismus: Eine Bestandsaufnahme* (Göttingen: Wallstein, 2010), a fine collection of essays on the Deutsches Museum during the Nazi period.

The founding of the Palais de la Découverte on the occasion of the Exposition internationale des arts et techniques dans la vie moderne in Paris in 1937 was colored by a political context in France inseparable from the worsening international relations of the 1930s. The point is made, with different emphases, in Jacqueline Eidelman, "The Cathedral of French Science: The Early Years of the 'Palais de la découverte,'" in Terry Shinn and Richard Whitley, eds., *Expository Science: Forms and Functions of Popularisation* (Sociology of the Sciences Yearbook 1985; Dordrecht, Boston, and Lancaster: D. Reidel, 1985), 195–207, and Charlotte Bigg and Andrée Bergeron, "D'ombres et de lumières: l'exposition de 1937 et les premières années du Palais de la découverte au prisme du transnational," *Revue germanique internationale* 21 (2015), 187–206. As an international setting in which politically motivated national interests were very much to the fore, the 1937 exhibition as a whole has begun to attract historians' attention. On the complex relation between German and French interests at the exhibition, Karen Fiss's *Grand Illusion: The Third Reich, the Paris Exposition, and the Cultural Seduction of France* (Chicago: Chicago University Press, 2009) has set a high standard, both visually and in its analysis.

On other international exhibitions of the 1930s and the national aspirations they evoked and promoted, the literature is extensive and growing. Robert Rydell's *World of Fairs: Century-of-Progress Exhibitions* (Chicago: University of Chicago Press, 1993) opens important perspectives on the American exhibitions of the Depression years and the politics that informed them. A more general work on the period is Robert H. Kargon, Karen Fiss, Morris Low, and Arthur P. Molella, *World's Fairs on the Eve of War: Science, Technology, and Modernity, 1937–1942* (Pittsburgh, PA: University of Pittsburgh Press, 2015), which includes chapters not only on the Paris exhibition of 1937 but also on exhibitions in Dusseldorf (1937) and New York (1939–1940) and two exhibitions, planned for Tokyo (1940) and Rome (1942), that did not take place.

Revealing though exhibitions are, in science the political tensions associated with the rise of the totalitarian regimes of the 1930s left their best-known mark on the enforced emigration of Jewish scientists from Hitler's Germany; on this, John Cornwell, *Hitler's Scientists: Science, War and the Devil's Pact* (London: Viking, 2003) offers a good introduction to a vast literature. But in Italy too, the careers of Jewish scientists were affected, though not significantly until the Fascist Manifesto of Race promulgated in Italy in 1938. Here, Roberto Maiocchi's works on the relations between science and the Fascist state are essential starting points; see in particular his *Scienza italiana e razzismo fascista* (Florence: La Nuova Italia, 1999) and *Gli scienziatai del Duce: Il ruolo dei ricercatori del CNR nella politica autarchica del fascismo* (Milan: Carocci, 2003).

Elsewhere totalitarian suspicion of internationally minded intellectuals had different consequences, as sources on the Soviet Union and Franco's Spain make plain. On the Soviet Union, Loren R. Graham's work remains fundamental. His *Science in Russia and the Soviet Union: A Short History* (Cambridge: Cambridge University Press, 1993) is a fine introduction, with a helpful bibliographical essay. On Spanish science under Franco, the essays in Amparo Gómez Rodríguez and Antonio Francisco Canales Serrano, eds., *Ciencia y fascismos: La ciencia Española* (Barcelona: Laertes, 2009) provide a good overview. While scope remains for a comprehensive synthetic account, a body of more finely focused studies of science in Franco's Spain is beginning to accumulate, including important work in English; see, for example, Antoni Malet, "José María Albareda (1902–1966) and the Formation of the Spanish Consejo Superior de Investigaciones Científicas," *Annals of Science* 66 (2009), 307–332; Agustí Nieto-Galan, "From Papers to Newspapers:

Miguel Masriera (1901–1981) and the Role of Science Popularization under the Franco Regime," *Science in Context* 26 (2013), 527–549 (with a valuable extended bibliography); and, on the centrality of civil engineers and agricultural scientists in the regime's attempts to promote a distinctive technological nationalism, Lino Camprubi, *Engineers and the Making of the Francoist Regime* (Cambridge, MA, and London: MIT Press, 2014). Recent perspectives on science under totalitarian regimes more generally are well presented in Amparo Gómez, Antonio Francisco Canales, and Brian Balmer, eds., *Science Policies and Twentieth-Century Dictatorships, Spain, Italy and Argentina* (Farnham and Burlington, VT: Ashgate, 2015), an important collection of essays, mainly on Italy and Spain but including two on Argentina between the military coup d'état in 1930 and the restoration of democracy in 1983.

The secondary literature reminds us repeatedly how many of the key debates that I discuss in *Science without Frontiers* have echoes in our own day. The result is a sense of continuity, though a continuity that, as historians, we have to view with caution. Akira Iriye's analyses of "cultural internationalism" (pursued by nongovernmental bodies, such as the ICIC and UNESCO, rather than by states) make the point. As Iriye argues, this form of internationalism assumed new forms after the Second World War, and it has continued to evolve as we move toward a world order that Iriye sees as structured around global networks of information and defined in cultural terms, rather than economic or geopolitical influence. Two books in particular, *Cultural Internationalism and World Order* (Baltimore, MD, and London: Johns Hopkins University Press, 1997) and *Global Community: The Role of International Organizations in the Making of the Contemporary World* (Berkeley, CA, and London: University of California Press, 2002), map out Iriye's vision. They convey his belief that, despite its many setbacks, cultural internationalism is destined to undermine perceptions of ourselves as citizens of a nation in favor of a "new globalism" that will rise above nationhood. That segment of history, of course, is still in the making. But it raises issues that would have engaged the optimists of an earlier age that I consider in this book.

INDEX